FIREWORKS
IN A
DARK
UNIVERSE

Other Related Titles from World Scientific

From the Sun to the Stars
by James B Kaler
ISBN: 978-981-3143-75-3
ISBN: 978-981-3143-24-1 (pbk)

Beyond the Galaxy: How Humanity Looked Beyond Our Milky Way and Discovered the Entire Universe
by Ethan Siegel
ISBN: 978-981-4667-23-4
ISBN: 978-981-4667-16-6 (pbk)

First Magnitude: A Book of the Bright Sky
by James B Kaler
ISBN: 978-981-4417-42-6

FIREWORKS
IN A
DARK
UNIVERSE

Amir Levinson

Tel Aviv University, Israel

World Scientific

NEW JERSEY · LONDON · SINGAPORE · BEIJING · SHANGHAI · HONG KONG · TAIPEI · CHENNAI · TOKYO

Published by

World Scientific Publishing Europe Ltd.

57 Shelton Street, Covent Garden, London WC2H 9HE

Head office: 5 Toh Tuck Link, Singapore 596224

USA office: 27 Warren Street, Suite 401-402, Hackensack, NJ 07601

Library of Congress Cataloging-in-Publication Data
Names: Levinson, Amir, author.
Title: Fireworks in a dark universe / by Amir Levinson (Tel Aviv University, Israel).
Description: Singapore ; Hackensack, NJ : World Scientific, [2018] |
    Includes bibliographical references and index.
Identifiers: LCCN 2018001919| ISBN 9781786345110 (hardcover ; alk. paper) |
    ISBN 1786345110 (hardcover ; alk. paper)
Subjects: LCSH: Astronomy--Popular works. | Cosmology--Popular works.
Classification: LCC QB44.3 .L48 2018 | DDC 523.1--dc23
LC record available at https://lccn.loc.gov/2018001919

British Library Cataloguing-in-Publication Data
A catalogue record for this book is available from the British Library.

Translation of manuscript from Hebrew to English: Susan Spira

For any available supplementary material, please visit
http://www.worldscientific.com/worldscibooks/10.1142/Q0150#t=suppl

Desk Editor: Ng Kah Fee

Typeset by Stallion Press
Email: enquiries@stallionpress.com

"Education is that which remains if one has forgotten everything
he learned in school"

— Anonymous, as quoted by Albert Einstein

Preface

The day began like any other quiet day in the expanses of the universe. Billions of light years away from Earth, hundreds of millions of star-speckled galaxies floated around in the silent eternal vastness, with only pale star light faintly illuminating the timeless darkness. Nothing indicated what was about to transpire. Suddenly, a powerful explosion shook the cosmic stillness; a huge star had collapsed to its death in one of the faraway galaxies, releasing a rush of powerful jets of energy and matter. Within a split second, the core of the collapsing star became a black hole surrounded by magnetized matter from which the jets erupted. The entire event lasted about 20 sec, during which a significant part of the gravitational energy was emitted as a tremendous burst of gamma radiation. About one billion years later, some of that same radiation reached Earth and was detected by detectors installed on research satellites of various space agencies. Measurements revealed that the total energy emitted during the 20 sec duration of the explosion was a billion times greater than the total energy emitted by the sun in the 100 years that preceded the event. For several months after the explosion was observed, telescopes on Earth and research satellites in space measured strong radio emission, light, and x-ray radiation coming from the same position — radiation that was emitted from the huge blast wave caused by the explosion of the star, which continued to propagate through the interstellar medium for a very long time.

This scenario is not at all fictional! Events such as this — the grandest fireworks show in the universe — take place at least once a day in the vastness of space and are documented using the most advanced, state-of-the-art means currently available to astronomers throughout the world. Such events amass to a large gamut of extreme and mysterious phenomena exhibited by the universe, whose power source is a compact object — a black hole, a neutron star, or a white dwarf — with such a strong gravitational force, that it may be described only using Einstein's theory of general relativity. Despite the huge diversity of these spectacles and the seemingly fundamental differences between them, it is gradually becoming clear that the mechanisms underlying them all are basically very similar. Intensive theoretical and observational research conducted by some of the world's leading scientists, using some of the most advanced technological means ever developed, have led over the years to the exposure and analysis of a variety of events such as the one described above. Researchers have also gained an understanding of the relationship between the most basic theories in modern physics, particularly the general theory of relativity and the quantum theory, and the behavioral patterns of the most extreme objects the universe presents.

Introduction

The impetus for writing this book was a series of popular science lectures I gave at several Israeli high schools as part of my volunteer activities. The meetings with young audiences in non-formal settings are an empowering experience for the lecturers no less than for the students. To my surprise, in these encounters, I discovered a large number of knowledge-thirsty students with varied areas of interest. Their insatiable curiosity and interesting questions, which often gave me new perspectives on different subjects, served as an important motivator in my decision to invest time and energy in writing this book.

It is my humble opinion that the objective of a science book that addresses an audience of non-scientist readers is not to teach a subject in depth, but to offer a general picture, and particularly to pique the reader's curiosity and trigger a desire to delve deeper. As a youngster, I was attracted to physics mainly thanks to bits of knowledge I acquired from the popular science books I read and the public lectures I attended, most of which I did not even understand. The things I was told then about quantum theory and relativity theory seemed to me to be mystical and piqued my wonder and great curiosity. The desire to make sense for myself of what really underlies those fragments of information is what ultimately (and unintentionally) led me to choose physics as my profession. One of my objectives, when deciding to write this book, was to open a window for the reader into a world of strange and interesting phenomena that the universe displays, and which have never before been discussed extensively in popular literature, to explain their connection to theories of modern physics, and to review the various experiments and observational means that different countries have developed over the years in order to investigate and study those phenomena. The book is intended for any person who is interested in the subject and it does not require any prior knowledge other than the general knowledge normally acquired in junior high school.

Writing a popular science book is not a simple task for a "professional physicist". The language physicists use to express ideas, and communicate with one another is the language of mathematics. This language, in addition to being universal, enables us (physicists) to express profound ideas in a concise manner, and more importantly, in a quantitative manner. It was for that reason that Isaac Newton toiled to develop a new branch of mathematics (Infinitesimal calculus) that would enable him to formulate his laws of mechanics and gravitational theory, which he developed in a definitive manner (Newton's famous book, published in 1687, called *The Mathematical Principles of Natural Philosophy*). But while for physicist, using mathematical language is the most natural way of explaining things, to an average person, mathematical formulations appear as an inexplicable collection of undecipherable symbols. The need to translate the ideas underlying the formulations into simple verbal explanations in intelligible language is the main difficulty encountered when writing a text intended for the general public. This is why, in this book, I attempted to build the text around photographs and illustrations, following the well-known cliché, "One picture is worth a thousand words". I hope that even if not every single reader completely understands every single explanation, the incorporation of illustrations and fascinating photographs taken by telescopes throughout the world and by satellites operated by different space agencies into the text, helps deliver the message.

The book is divided into three main parts. The first part presents the reader with a general background of astrophysics, and has three chapters. Chapter 1 discusses the principles of modern cosmology. Phenomena described in this chapter include, among others, the Big Bang, the formation of matter and forces, Hubble's discovery of the expanding universe, cosmic background radiation measurements and the information they reveal, and the discovery of dark matter and dark energy. Chapter 2 addresses the types of galaxies that exist in the universe and their composition, and Chapter 3 describes the structure and evolution of stars. The second part of the book has six chapters dedicated to a description of the principles of modern physics and modern astronomy. It begins with a review of the structure of matter, the discovery of antimatter, the development of quantum theory and the important role it plays in modern physics, types of forces in nature, and the importance of symmetry and symmetry breaking in physics. Finally, a description is given of the different types of radiation used today to study the universe and its secrets, and the ways in which those radiation types are measured. Some of the more important experiments located throughout the world and on satellites in space are described as well. The third part of the book comprising nine chapters presents the main topic: the variety of extreme phenomena observed in the universe that are related to three strange objects — black holes, neutron stars, and white dwarfs — in which quantum mechanics and general relativity play a major role. Readers will find a description of the main features of each of these objects, the phenomena for which each object is responsible, and a simple explanation of the mechanisms underlying each phenomenon.

1. Dramatic Developments in the 20ᵗʰ Century

In 1905, a German journal published an article entitled "On the Electrodynamics of Moving Bodies". In this paper, the author, a student named Albert Einstein, who was at the time working at the Bern patent office, formulated new laws of physics that were later called *the special theory of relativity*. The article, which was ground breaking in its perception of the concepts of time and space, was the third in a series of four articles that Einstein published in that prestigious journal that year (which is referred to as "the miraculous year").[1] In another article in the series, a brief, page and a half long essay, Einstein described a *gedanken* (thought) experiment he conducted as part of the theory of special relativity, that led to the unification of the concepts of mass and energy in the famous equation $E = mc^2$, whose dramatic, scientific, political, and social implications in the 100 years since, have exceeded those of any other human discovery. This equivalency between mass and energy, the

---

[1] The fifth article Einstein published that year, entitled "*A New Determination of Molecular Dimensions*" earned him a doctoral degree from the University of Zurich.

practical meaning of which is that mass can be converted to energy and energy can be converted to mass, led to the development of the atomic bomb and nuclear weapons, enables the operation of nuclear reactors, and as we shall see later, is the energy source of the Sun and other stars in the universe. Einstein's second article dealt mainly with the kinetic theory of gases (Brownian motion), while in the first article in that series, Einstein introduced the quantum model of light, which will be discussed in detail in Chapter 4, Section 3, and which provides an explanation for the photo-electric effect. In 1921, Einstein was awarded the Nobel Prize in Physics for this article (not for the development of the relativity theory as is commonly believed). And so began a new era in physics. This era was yet another stage in the accelerated scientific and technological renaissance that began in the 16th century with the Copernican Revolution, which manifested not only in the perception of planetary motion, but rather in scientific method, and which led to the establishment of chemistry, physics, life sciences, and geology as modern scientific disciplines. This revolution also led to the discovery of the circulatory system and the breathing mechanism, to the development of the continental drift model, to Charles Darwin's theory of the origin of species, to the great works by Galileo Galilei and Isaac Newton on gravity and the dynamics of bodies, to the invention of the steam engine and the rapid technological developments that followed, to the discovery of the chemical elements and Mendeleev's Periodic Table, and to the development at the end of the 18th century of the theory of electromagnetism, which eventually led to Einstein's great discoveries, described earlier.

In 1915, after six years of arduous research, Einstein succeeded in generalizing the concept of relativity to systems with gravitational fields. In an essay published a year later, entitled "The Foundation of the General Theory of Relativity", Einstein set out the principals of his new theory as well as the philosophical motives that led to its development. The climax of this essay was the formulation of equations later to become known as the *Einstein field equations*, which define the relationship between the distribution of matter and energy in space and the geometry of that space. This geometry, according to the general theory of relativity, is the expression of gravity as we feel it in our daily life. Every physical object moves in this spacetime along a curved trajectory that depends on the geometry of the spacetime, and so the object actually "feels" the gravitational force applied on it by matter around it.

Shortly after the theory of general relativity was published, a scientist named Carl Schwarzschild derived an exact solution to the Einstein field equations that described a most strange object whose properties were incomprehensible to scientists of that time. Several decades passed before leading physicists understood and fully appreciate the significance of Schwarzschild's solution, which Princeton physicist and one of the leading researchers of relativity theory and quantum theory, John Archibald Wheeler, later called a *black hole*. The property that sets black holes apart is that its surface gravity (or, alternatively, spacetime warping) is seemingly so strong as to prevent any entity, including light or any other form of radiation, from escaping out from it. Thus, the inevitable fate of any object or body that comes close to the boundary of a black hole is to be swallowed into it through a unidirectional membrane, ultimately leading to the inevitable encounter with the singularity of spacetime. The quantum theory, which developed concurrently with the development of the general theory of relativity, led to the understanding that another strange object may exist in the universe, a kind of huge and very dense atomic nucleus in which the force of gravity is balanced by the quantum pressure of the neutrons — charge-free particles in the atomic nucleus that act as a kind of stabilizer that prevents the nucleus from flying apart. This object, in which the theory of general relativity and the quantum theory play a central role, and which is called a *neutron star*, was finally discovered in 1967. Robert Oppenheimer, who later headed the famous *Manhattan Project*,[2] developed the first detailed theory of neutron stars.

The rapid development of astronomical instrumentation that began in the 1950s led to the collection of a great deal of evidence for the existence of black holes and neutron stars in the universe. We know today that these objects are created when a normal star ends its life and collapses in a strong explosion called a *supernova*. The mass of black holes and neutron stars created in such a way is similar to that of our Sun. These bodies are the driving force behind phenomena such as pulsars, magnetars, gamma

---

[2] *The Manhattan Project* is the code name the Americans gave the project whose goal was to develop the first atomic bomb during World War II.

ray bursts, and microquasars, which were discovered in the age of modern astronomy and will be described at length in the last part of this book. Much evidence, however, alludes to the existence in the universe of another kind of black holes called *giant black holes*. The mass of these ranges from one million to 10 billion times the mass of the Sun, and they are located in the center of galaxies. In fact, most of the galaxies in the universe, including our Milky Way, probably contain such a black hole. It is still not yet entirely clear how giant black holes are created; they may grow over time due to the swallowing of galactic matter. At times, some of these systems appear as objects that astronomers refer to by different names such as quasars and radio galaxies (which will be described later on as well). These systems emit a broad spectrum of radiation, from radio waves to gamma rays, and in some of them, matter can be seen moving at velocities that approach the speed of light. And here is a paradox: how can we "see" black holes when, indeed, we claim that everything, including light and radiation, is trapped within them. What we see is, in fact, the impact black holes have on their closest surroundings, those close enough to feel the immense gravitational force of the black hole, but still within safe range from the inevitable fall into its eternal depths, where time and space change shapes. The regions around black holes contain the largest particle accelerators in the universe — hundreds of times larger than the Large Hadron Collider (LHC) at CERN laboratories in Switzerland — and they are reserved an important role in our story.

Acknowledgements

When working on the book, I deliberated extensively over the choice of topics and how to present them. The many comments and wise advice I received from friends, family members, and colleagues provided me with a great deal of food for thought and helped me shape the final version of the book. I would like to especially thank Prof. Haggai Netzer and Prof. Meir Meidav of Tel Aviv University for agreeing to read early versions of the manuscript and contribute from their rich experience and great wisdom. The structure and order of topic presentation, as reflected in the final version, are the outcome of their experienced advice. I also wish to extend special thanks to my son, Rotem Levinson, for casting a different light on various points and urging me to include several topics in the book that were not on my original agenda.

I also wish to thank Prof. Avi Loeb of Harvard University, Prof. Avery Broderick of the University of Waterloo (Canada), Prof. Stephan Rosswog of Jacobs University (Bremen, Germany), and Prof. Miguel Aloy of the University of Valencia (Spain) for graciously providing me with high-quality images of computer simulations they performed.

Contents

First Episode

The Universe and All That Is In It

The menagerie of celestial bodies offers a record of the dynamical processes that lead to the formation and evolution of galaxies, stars, and other complex structures in our ever-expanding universe. This expansion began nearly 14 billion years ago with a "Big Bang", which gave rise to time, space, matter, and forces, and will continue, so it seems, forever-more. In this section, we briefly review the structure and contents of the universe.

Chapter 1

The History of the Universe

The universe is home to stars and galaxies, as well as dark matter, whose nature is still unknown, but whose effects are apparent. A divine spirit dwells in the universe, in the form of dark energy, whose properties are even less understood than those of the dark matter. And all of this splendor and glory basks in a huge bath of microwave radiation called the *cosmic background radiation*, an ancient remnant of the primordial universe. The exact proportions of the various components, as measured in astronomical experiments conducted over the past two decades, is about 4% visible matter, namely stars and galaxies as well as gas and radiation permeating the space between the galaxies; around 23% dark matter, and the rest — about 73% — is dark energy.

The branch of science that deals with the universe is called *cosmology* (from the Greek *kosmus* = world, *logia* = study of). The progress made in the theoretical understanding of gravitational systems following the publication of Einstein's general theory of relativity, alongside several key discoveries of the 20th century, led to the development of a detailed cosmological model that describes the evolution of the universe from its birth until the present time. This is the subject of the following account. But before we delve into the main topic, we should first put things into their proper historical perspective.

1. The Universe According to the Ancients

Humanity has been interested in the movement of the planets and moons from as far back as ancient times. As early as the 4th century BC, Aristotle proposed his *geocentric model* (*geo* = earth, *centrum* = center), according to which the planets move at a uniform speed in perfect circular orbits, whose common center is Earth. At that time, the Sun and Moon were considered to be planets, alongside Mercury, Venus, Mars, Jupiter, and Saturn. Beyond these was another sphere that included the fixed stars, which were always seen in the same place in the sky, and beyond all of those, was yet another sphere, external to all the others, that set all the other spheres into motion. This was Aristotle's version of the universe.

Aristotle's model, which was graced with Platonic simplicity, suffered from two main problems: the first was the fact that the planets' luminosity varies, which contradicts the assumption that they move in perfect circular orbits around Earth, and the second was that Mercury and Venus reverse their motion across the sky. These phenomena led astronomers to refine Aristotle's model by adding secondary spheres called *epicycles*. According to the improved model, each of the planets moves in its own little sphere (epicycle), whose center revolves around Earth, along a larger, circular path called a deferent (like a sphere within a sphere). Ptolemy, a Greek astronomer who lived in Alexandria in the 2nd century, described the more detailed model, which included additional modifications, in his book *Almagest*. Ptolemy's model entailed great complications in order to explain the observations, but it was nevertheless adopted by many of his contemporaries. Over the years, the Church, too, adopted Ptolemy's model (see Figure 1), as it conformed with the perception whereby Earth is the center of the universe, and the model took on interpretations of a religious nature.

The complexity and inelegance of Ptolmey's model made many scholars uncomfortable, but since it was adopted by the Church, nobody dared undermine it for over 1000 years. The change in this perception was brought about by Nicolaus Copernicus in the 16th century, when he proposed a *heliocentric model* (*helio* = Sun, *centrum* = center), in which the planets, including Earth, revolve around the Sun. Copernicus even claimed that Earth revolves around its own axis once a day. His rationale was that his model explains the observations in a simpler and more natural manner. It should be mentioned here that the idea that the Sun is the center of the universe was introduced as early as the 3rd century BC by Aristarchus of Samos. This idea, however, was almost completely ignored since it seemed contrary to intuition (how could Earth be moving if we cannot feel

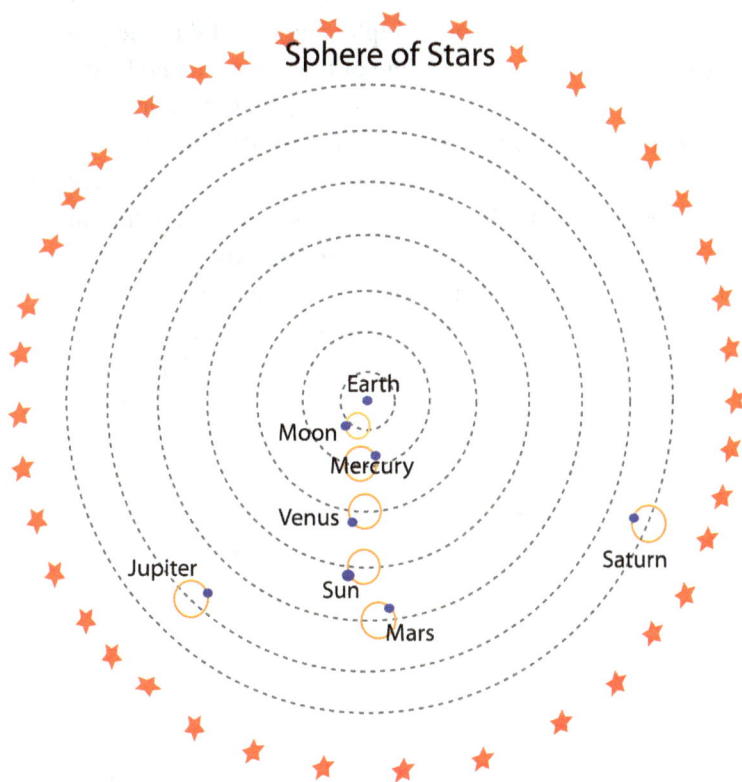

Figure 1. Ptolemy's model of the universe.

it move?) and also contrary to the prevalent philosophy at that time, according to which Earth is the center of the universe. The Copernican model predicted the correct order of the planets and even explained the seasons, and here too, the fixed stars were in fixed, unchanging positions.

Despite its success, the Copernican model suffered from similar problems to those that arose regarding Ptolmey's model, since it too assumed that the planets revolve in perfectly circular orbits. This problem was ultimately solved by a German named Johannes Kepler who served as an assistant to Tycho Brahe, who is considered to be one of the most important astronomers of the 16th century and is known for his many accurate observations that he conducted over the years. Based on Tycho's amazingly accurate observations, Kepler suggested that the planets travel in elliptical orbits with the Sun at one of the two foci, and he even discovered an empirical connection between a planet's distance from the Sun and its orbital period. Kepler ultimately formulated the principles of planetary motion in the form of three laws, which are named after him. The physical explanation of *Kepler's laws* was provided later on by Isaac Newton, based on the theory of gravity that he developed.

It is interesting to note that Tycho had reservations regarding Copernicus' heliocentric model, despite persuasive attempts by Johannes Kepler's, who was an avid supporter of this model. The main reason for this reservation was the fact that despite many attempts, Tycho failed to find any evidence of a *parallax*[1] of the fixed stars. We now know that the fixed stars are so distant from the Sun that their parallax is too small for Tycho to have measured it with the means available to him at the time. In order to explain his observations, Tycho proposed a model that is a kind of hybrid between the heliocentric approach and the geocentric approach. Despite Tycho's reservations, and thanks to his accurate observations, evidence supporting the heliocentric model increasingly accumulated.

Copernicus' work, and especially the way of scientific thinking he developed, had a significant impact on the later development of science, which ultimately led to the modern scientific thinking as it is practiced today. The way in which he perceived the world, which was revolutionary in his time, is (justifiably) referred to as the *Copernican Revolution*.

2. The Big Bang

The Big Bang is the name given to the modern theory of the universe. It is the current prevailing theory and is supported by numerous measurements conducted from the beginning of the 20th century to the present day. Ironically, the theory was named by one of its professed opponents and the developer of an alternative theory called the *Steady State theory*, British astronomer Sir Fred Hoyle, as an expression of his scorn. However, as

---

[1] A *parallax* (shift) is the change in the location of an object when observed from two different viewpoints. The motion of Earth around the Sun causes the relative positions of distant stars in the sky to change with the seasons. This effect was an important prediction of the heliocentric model.

support for the theory increased, the name stuck, whereas Hoyle's Steady State theory was abandoned completely.

Up until the beginning of the 20th century, the commonly accepted view was the Aristotelian view, according to which the universe is eternal and static, with no beginning and no end. Humanity's knowledge about the universe and its contents was, at that time, very limited; galaxies other than the Milky Way in which we live were not known to exist, and the prevailing thought was that the Milky Way was, in fact, the entire universe. Only in the 1920s, following the work of Edwin Hubble (after whom the Hubble Space Telescope was named) did the existence of additional galaxies become clear, leading to the realization that the Milky Way is only one of many galaxies in the universe. Hubble was also the one who discovered that the universe is expanding, one of the most important scientific discoveries ever, as will be described, in detail, later on.

The assumption that prevailed up until the beginning of the 20th century, namely that the universe is eternal and infinite, raised many difficult questions. One of the most famous of these questions, which was posed by Heinrich Wilhelm Olbers at the beginning of the 19th century and is known as *Olbers' Paradox*, is "Why is the sky so dark at night?". If indeed the universe is infinite and eternal, the light from the stars should accumulate infinitely and be seen equally in all parts of the universe. The inevitable conclusion from Olbers' Paradox is that the universe has a finite volume or that it is not eternal, in other words, it was created a finite time ago. The Big Bang theory indeed provides an answer to that question.

According to the Big Bang theory, the universe was born 13.7 billion years ago as a singular point (a point at which the density of matter and the curvature of space are infinite) that rapidly expanded and is still continuing to expand. During the expansion of the universe, elementary particles, radiation, atoms, and finally the galaxies and stars were formed. The question "What existed before the Big Bang?" is not relevant in this theory, since time and space themselves were only created at the time of the Big Bang itself.[2] Before continuing with our description, we should note that the laws of physics, as they are known today, cannot describe

---

[2] Not everyone is in agreement on the statement that time was "created" in the Big Bang. Sean Carroll discusses this in his book, *From Eternity to Here*, Dutton, 2010.

what took place in the initial split second during which time and space were created, and the forces of nature were still unified. They do, however, provide a good description of the expansion of the universe from the stage at which the gravitational force and the strong force separated from the other forces, a stage that began about 10^{-11} seconds after the creation of the universe in the Big Bang.

In fact, the expansion of the universe is described by Einstein's general theory of relativity, although Einstein himself was not aware of it at the time. Einstein believed that the universe is static, in other words that the distances between stars and other objects in the universe always remain constant. This idea stemmed from the observational distinction according to which distant stars do not change their position in the sky, and from the then-prevalent belief that the Milky Way, which contains these stars, is the entire universe. Einstein himself was the first to develop a model of the universe based on his theory, through which he calculated the radius of the universe as a function of the density of matter within it. However, he encountered a serious problem. The equations of general relativity (now called the Einstein field equations) imply that the universe must either collapse or expand, and therefore cannot be static. To resolve this conundrum, Einstein added an additional term to the equations, the *cosmological constant*, whose effect is like that of an anti-gravity force, to offset the attraction caused by the gravitational force of matter in space. Thus, Einstein obtained a model of the universe in which the average density of matter does not change over time, and space while indeed limitless, is not infinite but closed. Many years later, after Hubble's discovery regarding the expansion of the universe, Einstein wrote that the concept of a cosmological constant was the "biggest blunder" of his life. Ironically, observations conducted over the past two decades have shown that the existence of a cosmological constant, nowadays referred to as *dark energy*, is an inevitable consequence of the data.

An important step in the evolution of modern cosmology was the work done by Alexander Friedmann, published in 1922. Friedmann found an entire family of solutions to Einstein's equations, some of which describe an expanding universe that contains matter and others, a contracting universe. He also showed that there are solutions that do not require a cosmological constant. In order to explain why the universe has not yet collapsed, Friedmann had to assume that it was given some kind of forceful initial thrust, following which it began expanding. He even demonstrated three options that exist regarding the evolution of the universe, which depend on the amount of matter the universe contains (or more precisely, the average density of this matter). If this density is high enough, it will cause strong gravitational attraction that will halt the expansion, and will gradually cause the universe to contract until it collapses completely. This solution is called a *closed universe*. If, on the other hand, the density is too low, the gravitational force will slow down the expansion of the universe only slightly, but will not be able to stop it, and the universe will continue to expand forever at a constant pace. This solution describes an *open universe*. The third option is if the density has some critical value for which the universe will neither collapse inward nor continue to expand forever. This is a *flat universe*. As their names indicate, these three solutions, which Friedmann found, correspond to three different spacetime geometries, as shown in Figure 2.

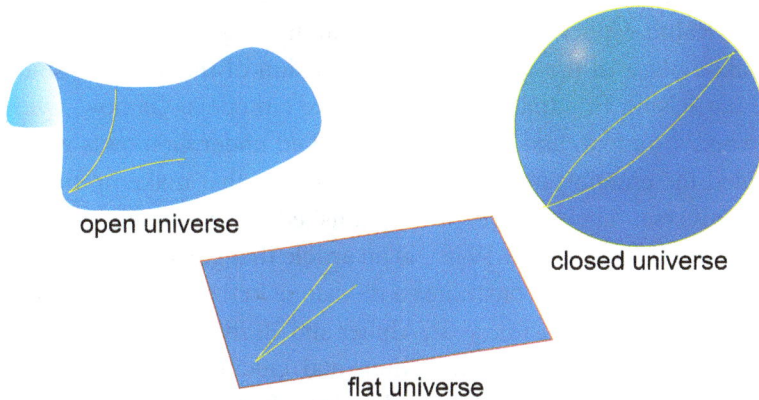

Figure 2. The geometric properties of each of Friedmann's three universes are demonstrated by the behavior of light rays (yellow lines): In a closed universe, two rays that are emitted from a source will ultimately meet; in a flat universe, the angle between the rays will remain constant throughout their entire travel; whereas in an open universe, the rays will diverge, or in other words, the angle between the rays will increase along their course. This is equivalent to the fact that in a closed universe, the sum of angles in a triangle is greater than 180°, in a flat universe the sum equals 180°, and in an open universe the sum is smaller than 180°.

Some five years later, a cosmologist named Georges Lemaître, who was a Belgian priest as well, published his own models of the universe, which were also based on Einstein's equations without a cosmological constant. In fact, Lemaître had re-derived Friedmann's solutions, although he was not aware of them at the time. However, Lemaître was not satisfied with simply analyzing the implications of the model on the future of the universe; he attempted to understand the past, as well. An analysis of the solutions led him to the conclusion that if the universe is currently expanding, then it must have been smaller and smaller, and denser and denser, the further back in time one goes. Lemaître suggested that the universe began at a single, very dense, high temperature point, which he called the *primeval atom*. According to Lemaître's scenario, the universe was created when this primeval atom, which contains everything, decomposed suddenly, similar to the way radioactive atoms decompose, and began expanding. This was, in fact, the first version of the Big Bang theory.

Friedmann's and Lemaître's pioneering works, followed by Howard Robertson and Arthur Walker who refined the model, constitute the foundations of modern cosmology and the Big Bang theory. Later on, George Gamow, Ralph Alpher, and Robert Herman investigated the processes that have taken place in the course of the expansion of the universe since the Big Bang. Using Friedmann's and Lemaître's models, Gamow, who was Friedmann's student, and Alpher, who worked under Gamow's guidance, explained the creation of the chemical elements that make up the matter in the universe. Their results were published in the prestigious journal *Physical Review* on 1 April 1948[3], in an article that gained great fame, and in the opinions of many constitutes a milestone in the establishment of the Big Bang theory.[4] That same year, Alpher and Herman predicted the existence of cosmic background radiation, and calculated its characteristic frequency. But it was only in 1965 that Arno Penzias and Robert Wilson actually discovered this radiation by accident. The discovery of cosmic background radiation, which will be discussed extensively later on, was

---

[3] "The Origin of Chemical Elements", *Physical Review*, 73(7), 803–804.
[4] Gamow jestfully added the name of his friend Hans Bethe, future Nobel Prize laureate and head of the Manhattan Project's theoretical division, as co-author of the paper so as to obtain the sequence of names Alpher, Bethe, and Gamow, whose sound resembles the first three letters of the Greek alphabet, alpha, beta, and gamma.

the "smoking gun" of the Big Bang theory, and put the final nail in the coffin of Fred Hoyle's Steady State theory.

When the cosmological constant is added to Friedmann's models, solutions may be obtained that describe a static universe and even an expanding universe, whose expansion rate not only does not decrease, but even increases with time. For years, scientists believed that such a phenomenon was not plausible, and that no cosmological constant exists in nature. However, as mentioned earlier, data collected in the past 20 years from various observations have proved that the expansion of the universe is indeed accelerating, and measurement of this acceleration lead to the conclusion that the cosmological constant (which is also known as dark energy) constitutes over 70% of the universe's energy content.

3. The Chronology of the Universe

The chronicles of the universe from its infancy to the present day may be described based on Friedmann's and Lemaître's model and using the known laws of physics. The birth itself, however, cannot yet be described using existing physics, mainly because a theory of quantum gravity has not yet been formulated successfully.

In the first split second after its creation, the universe was very hot and dense. The temperature was so high, that even atoms and atomic nuclei could not exist, since they melted immediately upon formation, just like iron cannot solidify in the furnaces of iron foundries but can exist only as gas or liquid. At this time, the universe contained a mixture of elementary particles called *quarks* and *gluons*, which are the components of atomic nuclei. (Quarks and gluons are discussed more in Chapters 4 and 12.) As the universe continued to expand, its density and temperature continued to decrease. At some point in time, when the temperature and density of the universe were low enough, quarks combined to form *protons and neutrons*, and immediately following that they also formed the nuclei of the first atoms. The size of the visible universe at that time did not exceed that of our solar system. The only elements created were lightweight elements, mainly hydrogen and helium nuclei, since the creation of heavier nuclei, such as iron, required more time than was possible under the conditions that prevailed at that time. The heavy elements were created later, within the cores of stars and through cosmic explosions that occur upon the death of stars, as will be discussed later on. Exact measurements of the gas composition of the young universe, which were compared with theoretical calculations, provide the third clear-cut proof (along with measurements of the expansion of the universe and the discovery of cosmic background radiation) of the correctness of the Big Bang theory.

Immediately after the formation of the first elements, the temperature of the universe was still too high for electrons to bond to the atomic nuclei to form atoms (a process referred to as *recombination*). At that stage, the universe contained ionized atomic gas in which the electrons were not bonded to the atomic nuclei, but rather moved about freely. Radiation, which was emitted due to the intense heat, was trapped in the ionized gas. Approximately 380,000 years after the initial explosion, the temperature dropped to below 3,000°C and the recombination process began. Within a very short period, the atomic nuclei captured the free electrons, and the gas went from being an ionized gas to being a neutral one containing hydrogen, helium, and additional lightweight elements. This is when the stage known, in the history of the universe, as the *dark ages*, began; stars and galaxies had not yet been created, and so there were no light sources to illuminate the abysmal darkness that prevailed in the universe.

As the universe continued to expand, it continued to cool. The distribution of gas in the universe was not completely uniform, and quantum fluctuations that remained from the early stage caused certain regions in the universe to be slightly denser than others. These deviations were, however, very small. Now, when the universe was sufficiently cool, the denser areas began collapsing due to their own gravitational force, and gaseous clouds, galaxies, stars, and other structures we are familiar with today began forming. Because the initial fluctuations in gas density were so small, the formation process of the galaxies and other structures lasted a relatively long time. According to current estimates, the universe was 30 million years old when the dark ages ended and the first stars were created, and it was 100 million years old when the first galaxies began forming.

4. Edwin Hubble's Great Discovery

Up until the 1920s, opinions within the astronomical community differed regarding the nature of spiral nebulae that were clearly discernable through telescopes available at that time. Many believed that they were small bodies positioned within the Milky Way, while others claimed that they were independent galaxies located outside of the Milky Way. To settle this debate, the distance of the observed nebulae had to be measured.

Measuring the distance of astronomical objects is one of the most central challenges in astronomy. How does one measure the distance to a distant star or another galaxy? We cannot simply stretch out a measuring tape between two points in the universe. Immense resources have been invested over the years in the search for methods that enable accurate measurement of distances to various objects in the universe. One of the most accepted methods is to deduce the distance to a certain celestial object by measuring the intensity of light that it emits, using the known fact that the brightness of a given source is inversely proportional to the square of the distance from that source. This relationship means that the flux of light measured from a light source of given luminosity that is located at a certain distance from Earth will be four times greater than the flux of light measured from an identical source located at twice the distance. In order to use this method, however, the luminosity of the source whose distance we wish to measure must be known. In principle, if a group of stars or other astronomical objects with identical and known luminosity was available, those objects could be used to measure distances. In astronomical jargon, such a group of objects is called *standard candles*. In 1908, American astronomer Henrietta Leavitt identified a group of stars that serves as standard candles. Leavitt was studying a type of stars called *Cepheid variable stars*, which are giant stars whose brightness varies periodically, alternatingly bright and dim. During the course of her work, Leavitt found a connection between the absolute brightness of a Cepheid and its period: the brighter the star, the longer its period. This relationship enabled researchers to deduce the brightness of a Cepheid star by measuring the period of the light it emits.

Henrietta Leavitt's discovery provided a means of measuring distances in the universe; based on the period of a star's light curve and its observed brightness, its distance from Earth could be calculated. Indeed, shortly after Leavitt published her findings, astronomers began mapping the size of the Milky Way using Cepheid stars located in various regions of the galaxy.

Hubble's great discovery, too, was possible thanks to those same varying Cepheids. During a dark night in October 1923, while Hubble was viewing the Andromeda Nebula through a telescope situated on Mount Wilson, California, he discovered a Cepheid star within the nebula. This discovery enabled Hubble to calculate, for the first time ever, the distance between Andromeda and Earth. The calculation showed that Andromeda is 900,000 light years away, which is a much greater distance than the observed size of the Milky Way Galaxy. Thus, Hubble proved that Andromeda is an external galaxy rather than a small nebula located within the Milky Way, as many had believed until then. This measurement, in fact, ended the debate regarding the nature of the spiral nebulae observed and led to the widespread recognition within the astronomical community that the Milky Way Galaxy is only one of many galaxies in the universe. It is currently estimated that there are about 200 billion galaxies in the visible universe. The discovery that the universe contains additional galaxies was of great importance at the time, but as we shall presently see, it was only the preliminary stage of a much more earth-shattering discovery.

About 10 years prior to Hubble's discovery, American astronomer Vesto Slipher showed that the light emitted from many spiral nebulae is shifted to the red end of the spectrum. At that time, techniques had already been developed to separate light into its components (a method called *spectroscopy*), and the color composition of stellar light, which reflects the stars' chemical composition and constitutes a kind of fingerprint, was known. Slipher found that for many nebulae, the color composition of the stellar light they emit was not exactly as expected, but slightly redder. More precisely, each of the colors of the spectrum observed was shifted towards the red to an identical extent. As we explain in the second part of this book, each color corresponds to a certain wavelength; red corresponds to a long wavelength, whereas blue corresponds to a short wavelength. The significance of the redshift Slipher measured is that each of the waves the nebula emitted had a longer wavelength than was expected. At that time, scientists already knew that such a change in wavelength was the result of the movement of the wave source, an effect known as the *Doppler Effect* (see Box 1). Thus, the conclusion from Slipher's measurements was that the nebulae that display a redshift are moving away from us. The reason for this movement, however, was not altogether clear, and the distance to the nebulae was still unknown.

Box 1. Redshift and the Doppler Effect

The Doppler Effect refers to the change in the observed frequency of a wave caused by the movement of the wave source relative to the observer. The effect, which is named after Austrian mathematician Christian Andreas Doppler who discovered it in 1842, can be observed in all kinds of waves: sound waves, electromagnetic waves, gravitational waves, and so on. When the source is moving towards the observer, the wave frequency increases (and the wavelength decreases), and when it is moving away from the observers, the wave frequency decreases. A common example of the Doppler Effect is the change in pitch of a honking car, from a high-frequency tone when it is travelling towards us, to a low-frequency tone after it has passed and is travelling away from us.

In the case of visible light, movement of the light source towards the observer will cause colors to appear bluer, a phenomenon known as blueshift, whereas movement in the opposite direction will cause a redshift.

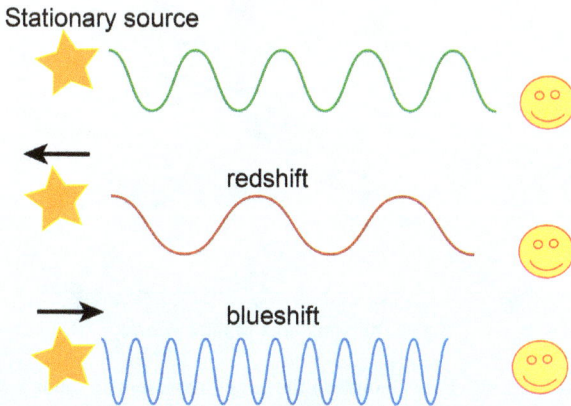

Stationary source

redshift

blueshift

The Doppler Effect has applications in medicine (in cardiac tests, for example), flow velocity measurements, the military, weather forecasting (Doppler radar), and astronomy, to name a few. In astronomy, the Doppler Effect is used to measure the period of binary stars, to discover planets, to conduct cosmological measurements such as the distance to very distant objects, to measure the movement of pulsars, and more. The Doppler Effect also helped establish the fact that the universe is expanding.

Hubble was familiar with Slipher's results. Using Cepheid stars, Hubble measured the distances to those nebulae (which were already known to be galaxies) for which a redshift was observed. He discovered a very interesting relationship: the farther the galaxy was from Earth, the greater the redshift of its light spectrum, and so the faster its movement away from Earth. Furthermore, the velocity of the galaxies moving away was directly proportional to their distance from Earth. Hubble also measured the proportionality constant, which was later called the *Hubble Constant*, a fundamental constant in the Big Bang model (see Figure 3).

The empirical relationship Hubble discovered fitted Friedmann's and Lemaître's predictions about the expansion of the universe like a glove. In fact, the apparent movement of galaxies away from Earth does not result from their true motion, but rather from the expansion of spacetime itself.[5] To demonstrate this, imagine the universe as the surface of an inflating

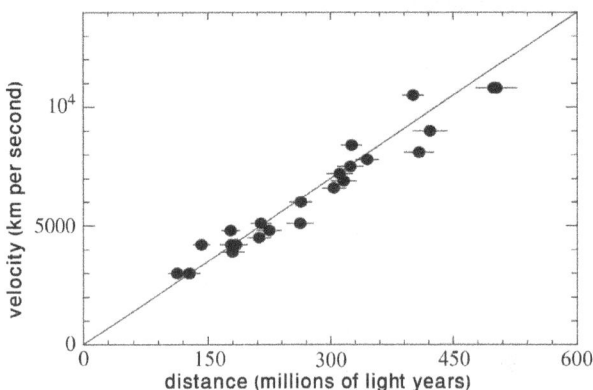

Figure 3. Hubble's Law: Every point on the graph represents a galaxy whose distance and velocity have been measured. The horizontal axis denotes the distance from Earth in millions of light years, and the vertical axis indicates the velocity at which the galaxy is moving away from Earth in km/s. As the straight line passing through the points emphasizes, the relationship between velocity and distance is linear. The slope of the line is the Hubble Constant.

---

[5] To be completely precise, the statement that the redshift that Hubble measured results from the Doppler Effect is not accurate. In fact, the shift is due to the expansion of spacetime itself rather than to the fact that the source is in motion. There is, however, a connection between the two, and the Doppler Effect explanation of the phenomenon has become commonly accepted.

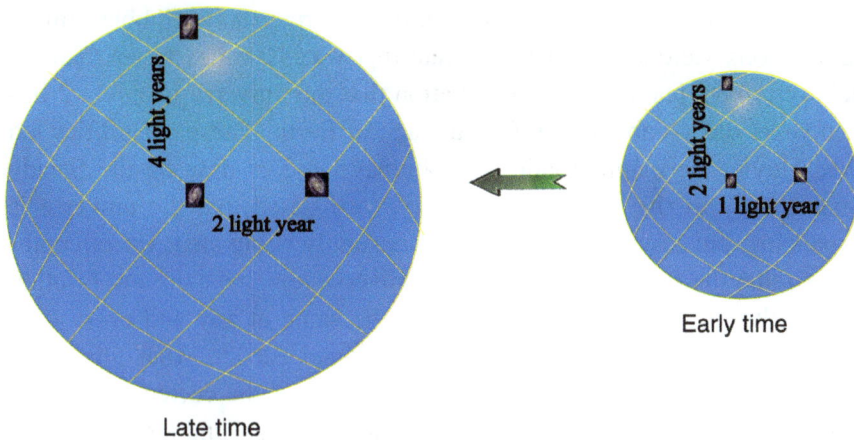

Figure 4. Illustration of an expanding universe: The surface of an inflating sphere represents the expanding universe. The yellow lines represent a universal coordinate system. The figure shows the universe in two states that correspond to two different times. As is evident, the distances between the three galaxies in the universe increase in direct proportion to the radius of the sphere, but their location on the yellow coordinate system remains unchanged.

balloon. All objects in this universe exist only on the balloon's surface and are unaware of anything else that exists in other dimensions (in other words, inside or outside of the balloon). If we draw dots on the surface of the balloon, such that each dot represents a galaxy, then it is clear that as the balloon inflates, the distance between the dots will increase, while the dots themselves remain fixed in place relative to the balloon surface (see Figure 4). In addition, since the distance between any two dots increases linearly with the radius of the balloon, it is obvious that their relative velocity also increases linearly with the increase in the distance between them. In other words, Hubble's law applies to the dots on the balloon surface.

Here is the place to note that Hubble's discovery, despite being remarkably consistent with the predictions of the Big Bang theory, did not yet constitute decisive support to the theory. In fact, the perception that Lemaître adopted, according to which the universe had a beginning, raised objections among many astronomers, and some even regarded the universe's birth scenario as a religious idea that stemmed from Lemaître's faith (Lemaître was, as mentioned, a priest).[6] Furthermore, Hubble's initial calculations yielded an expansion rate that was 10 times faster than we believe today,[7] leading to the estimation that the universe is only 1.2 billion years old — about one-third the age of Earth. These absurdities led Fred Hoyle, Hermann Bondi, and Thomas Gold to develop the Steady State theory in the 1940s. This theory, too, stated that the universe is expanding, but assumed that matter is constantly being created from nothing, so that the average density of the universe remains constant. In other words, the universe is and was exactly the same at any and every time (hence the name, *steady state*). As mentioned earlier, what ultimately thoroughly convinced the scientific community of the correctness of the Big Bang theory was the discovery of cosmic background radiation.

5. Cosmic Background Radiation — Signals from the Ancient Universe

In 1965, Arno Penzias and Robert Wilson, two physicists from the American phone company Bell, built a new radio antenna for their astronomy and satellite communications experiments. The antenna's range of sensitivity included the microwave range, which is the range of radio waves used in cellular phones to transmit and receive signals and in microwave ovens to heat food. While using the antenna, Penzias and Wilson measured a strange static noise that came from all directions of the sky. At first, they thought the noise was interference transmitted from the nearby city of New York, but this possible explanation was soon rejected. After rejecting every other possible cause of the noise that they could think of, they became convinced that the noise came from some unidentified astronomical source. Penzias and Wilson then conferred with a group of senior astronomers from Princeton University and came to the conclusion that the noise was radiation created during the expansion of the universe — cosmic background radiation — whose existence was predicted back in 1948 by Ralph Alpher and Robert Herman as part of the Big Bang theory.

---

[6] The Catholic Church, under the leadership of Pope Pius the 12th, officially adopted the Big Bang theory in 1951.

[7] The error, which resulted from a confusion between Cepheids and another kind of varying stars, was corrected later on.

In the years since Penzias and Wilson's experiment, the scientific community, and particularly the American space agency NASA, invested immense resources in studying the cosmic background radiation, which, as we shall presently see, carries valuable information on the nature of the universe. Many experiments were constructed especially for that purpose, including several space observatories, COBE, WMAP, and Planck, whose results enabled to measure the fundamental cosmological constants, and which together with the supernova experiment, which shall be described later on, revealed the existence of dark energy and unleashed a small revolution in the way we perceive the universe.

What then is the source of the cosmic background radiation? In Section 3, we mentioned that prior to the age of recombination, the universe contained ionized gas. Under such conditions, gas emits and scatters radiation effectively (indeed, this is the main source of radiation of the Sun and of other stars). The high density that prevailed in the universe at that time trapped the radiation emitted by the ionized gas. Immediately at the end of the recombination era, after all of the ionized gas had become neutral atomic gas, the radiation and gas separated. All at once, the universe became transparent to this radiation, because the electrons were now bonded to atoms, and could not scatter or emit radiation. From there on, the gas and radiation continued to develop separately. The more the universe continued to expand, the cooler and cooler became both the gas and the radiation. Decreasing the temperature of radiation manifests in a change in wavelength, and so the cooling radiation went from being ultraviolet to being visible light radiation, to infrared radiation, and finally becoming the microwave radiation that Penzias and Wilson detected. Nowadays, the temperature of the radiation is about 3 K (which is equivalent to about −270°C). Since this radiation is present everywhere in the universe, it is seen in any direction of the sky we look. It may be said that the entire universe bathes in this radiation, which is a remnant of what occurred before the dark ages of the universe and even before the age of recombination. The presence of the cosmic background radiation in the present time universe is an important confirmation that the universe was, in the past, both hotter and denser, a fact that only the Big Bang theory, and no other theory, including the Steady State theory, can explain.

The cosmic background radiation contains valuable information on the structure of the universe. In general, the radiation characteristics such as temperature and intensity are uniform across the sky. That means that in any direction we look, the values we measure will be the same. Yet there are, nevertheless, minute deviations, of a magnitude of 1 to 100,000, in these characteristics that reflect spatial changes that occurred in the density and temperature of the gas before the age of recombination, when the universe was only 380,000 years old. The origin of these perturbations is not entirely clear. Some models attribute them to quantum fluctuations in very early times, approximately 10^{-35} seconds after the universe was created, but there are other ideas as well. Be their origin as it may, thanks to these fluctuations, the stars and galaxies were ultimately created, as was life on Earth. Accurate measurements of the cosmic background radiation's characteristics provide a quantitative basis for the theory of structure formation in the universe and for the hypothesis of the existence of dark matter. Without dark matter, much greater deviations than those measured by the COBE, WMAP, and Planck satellites would have been needed to explain the creation of the stars and galaxies. Here is the place to mention that alternative theories have been proposed, which modify Einstein's general theory of relativity, and which instead of dark matter, assume a correction of the gravitational force on large scales. These theories have, at present, only a few supporters.

The temperature fluctuations in the cosmic background radiation (the spots on Figure 5(b)) are not distributed randomly, as it seemingly appears. Upon meticulous examination, a kind of wave-like pattern emerges. This pattern stems from sound waves that were created in the universe before the age of recombination, just like sound waves are formed in the air when we strum a guitar or tickle the ivories. Sound waves represent changes in the density and temperature (thus, also in the pressure) of the medium in which they travel. For instance, when we clap our hands, the air compresses and a disturbance is created in the surrounding medium. This disturbance expands through the air, from the focal point outward. When it reaches the listener's ear, the changes in the compression of the air cause the tympanic membrane in the ear to vibrate. Sound waves in the universe cause the creation of hotter and cooler regions in the ionized gas. Since the radiation is trapped within the gas at

(a)

(b)

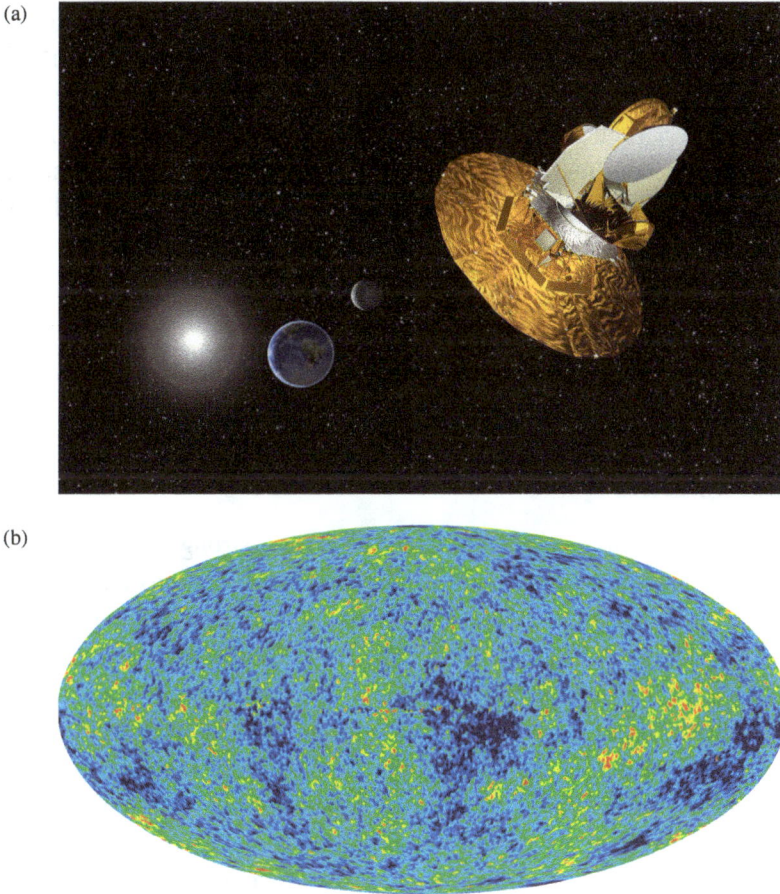

Figure 5. All sky map of the cosmic background radiation, as measured by the WMAP observatory (a). The spots in (b) are measurements of the radiation temperature variations across the sky. The red spots indicate hot regions, while the blue spots indicate cold regions. The extent of temperature variations between the red and blue spots is 1:100,000 of the average *temperature*. NASA / WMAP Science Team.

this stage, these temperature changes cause changes in the intensity of the radiation according to the same pattern — in hotter regions the radiation intensity is higher, and in cooler regions, it is lower. Immediately at the end of the recombination era, when the universe suddenly became transparent, the radiation separated from the matter, but the interference pattern caused by the sound waves remained imprinted on it forever.

The variation pattern in the radiation enables scientists to more accurately deduce the age of the universe, the geometry of the universe, the quantity of visible matter (the matter that makes up the stars, the galaxies, and the gas between them), the initial conditions that prevailed in the very early universe, and more. These measurements revealed that of the three possible geometries in Friedmann's model, the universe "decided" to be flat. It also became clear, as mentioned earlier, that visible matter constitutes only 4% of the universe's contents, and that the rest is dark matter and dark energy. These results also support the modern version of the Big Bang theory, which will be discussed later on, into which an early "inflationary phase" was also introduced.

6. Dark Matter

In the case of regular stars, including the Sun, a strong correlation exists between the intensity of the light emitted by the star and its mass; specifically, the greater the star's mass, the more light it emits and the brighter it appears. Using this relationship, the mass of any star may be deduced by measuring its luminosity. Since a galaxy is a collection of stars, the light emitted by a galaxy is the sum of the light emitted by its individual stars, and the mass of the galaxy (which equals the sum of the masses of all of the stars in the galaxy) may be deduced by measuring its luminosity. If the mass of bodies in the universe is known, then it is possible to predict how their gravitational force will affect other bodies. For example, by measuring the mass of the Sun, it is easy, using Newton's laws, to calculate the theoretical speed of revolution of Earth and the other planets. Indeed, the measured velocity is very close to the theoretical value. Similarly, by measuring the intensity of light emitted from various locations in any given galaxy, it is possible to deduce the distribution of its mass and hence predict the velocity of the stars in that galaxy.

In 1933, an astronomer named Franz Zwicky from the California Institute of Technology measured the velocity of galaxies in a *galaxy cluster* called the Coma Cluster. A cluster of galaxies is a collection of galaxies that are bound together by their mutual gravitational force permeated by a thin gas. A typical cluster may contain anywhere from several galaxies to thousands of galaxies, and may be up to 10 million light years in size. By measuring the velocities of the galaxies in the cluster as well as the intensity of the light they emit, Zwicky deduced that the mass that causes their motion must be approximately 400 times greater than the "visible" mass of the galaxies that make up the cluster. This phenomenon was called the *"missing mass problem"*, but since no additional instances of this phenomenon were found in the decades following Zwicky's discovery, this peculiar issue was forgotten and forsaken. The missing mass problem returned to center stage in the early 1970s, after a young astronomer named Vera Rubin introduced evidence of the missing mass in spiral galaxies (galaxies with a flat disc structure, like that of the Milky Way). Using innovative equipment, Rubin succeeded in measuring the velocity profile of stars in a few relatively close galaxies (in other words, she measured the circular velocity at various radii from the center of the galaxy). Rubin's measurements revealed that the velocity of stars located at a great distance from the center of the galaxy is much higher than that predicted by Newton's laws based on the light emitted by the galaxy. The simplest explanation of Rubin's data was to assume that the galaxy exist in a kind of halo of dark matter, which does not emit light or any other kind of radiation and whose mass is about 10 times the mass of the galaxy itself. Another possibility was that Newton's laws are not valid for such systems, and that the theory of gravity requires certain modifications. However, no satisfactory alternative theory has yet been found and, to date, not many researchers support this possibility; the majority prefers the hypothesis that dark matter does exist.

Since the publication of Rubin's findings, additional systems have been discovered in which mass is missing. At the same time, it has also become clear that, as mentioned before, the existence of dark matter is required in the Big Bang theory in order to explain certain measurements of the cosmic background radiation obtained by WMAP and in other experiments.

So what is this dark matter whose existence we deduce, but that we cannot see? The scientific literature discusses two main possibilities. The first is that dark matter consists of massive, non-radiating objects, such as small black holes, neutron stars, certain kinds of low-mass stars such as brown dwarfs, that have cooled down to an extent that prevents the detection of the radiation they emit, and so on. Since such objects are made up of regular atoms or atomic nuclei, this kind of dark matter was named *baryonic dark matter* (because it is composed of protons and neutrons, which are members of the baryon family). Attempts to find objects of this kind have been futile, and dedicated observations conducted especially for this purpose actually proved that such objects are not sufficiently abundant so as to constitute the dark matter. The second possibility proposed is that dark matter consists mainly of neutrino particles, or alternatively, exotic kinds of elementary particles, such as axions and super-symmetric particles, whose existence is theoretically possible, but which have not yet been discovered. Attempts to detect these particles using the new hadron accelerator, the LHC, are in progress. Although it is not clear, for now, what kind of particles (if at all) they are, most researchers are in agreement regarding one thing: they cannot carry electrical charge.

The dark matter enigma is still far from being solved, and continues to occupy the minds and time of the best physicists in the world.

7. Dark Energy

If dark matter is said to be elusive, it is all the more so when it comes to dark energy. Why must we assume the existence of dark energy in the first place? Well, if the universe consisted only of matter with regular mass or energy, including visible matter and dark matter, then its own gravitational force would resist the expansion of the universe. In such a case, we would expect to measure a deceleration in the expansion rate. In other words, even though the universe continues to expand, the expansion speed should decrease over time. This situation resembles a stone that is thrust upward to the sky. Immediately after being launched, the stone travels up, but its velocity decreases over time. If the initial thrust is not strong enough, the stone will eventually stop and fall back down. If, on the other hand, the stone's initial velocity is high enough, and more precisely, higher than the escape velocity from Earth, it will continue traveling upward forever, but its velocity will decrease over time, until it reaches a constant velocity at a very large distance from Earth. In no case will the velocity of the stone increase, unless it was equipped with a rocket engine that operates while in flight and continues to accelerate the stone (provided it is strong enough to overcome gravity). In this example, the instant in which the stone was thrust into the sky represents the Big Bang, when hidden forces "thrust" matter outwards into the universe. The self-attraction of the matter particles in the universe is similar to the force of gravity that acts upon the stone and causes it to decelerate.

In reality, the expansion of the universe was found to be accelerating, leading to the conclusion that some entity is opposing gravity and accelerating the mass in the universe (the "rocket engine"). That entity is dark energy. Dark energy has a kind of negative gravity,[8] whose action opposes the regular force of gravity and causes the universe to accelerate.

What evidence is there that the universe is accelerating? One way to tell how the universe is behaving is to measure the relation between the distance of very distant cosmological objects and their redshift, just like Hubble did. For objects that are not too distant on a cosmological scale, like the Cepheid stars Hubble observed, the effect is very small and cannot be discerned using common astronomical methods. In order to determine, with certainty, whether the universe is accelerating or not, such measurements must be made on much more distant objects. As explained above, direct measurements of such distances requires standard candles, and these must be especially luminous, tens of thousands of times more luminous than the stars that Hubble used, so that they may be discernible at the desired distances. Extended and in-depth research revealed that a certain kind of cosmic explosions called *type Ia supernovae*, which will be discussed in detail in Chapter 17, are a group of especially luminous standard candles that may be detected even at monumental distances. A special project, in which astronomers from all around the world participated, set out to locate such explosions in a methodical manner, and classify them according to their redshift. In 1998, after sufficient data were collected and it was possible to construct a kind of Hubble diagram for type Ia supernovae, the first piece of evidence of the acceleration of the universe was obtained. The three scientists who led most of these efforts were awarded a Nobel Prize for their work in 2011.

---

[8] A more accurate term is "negative pressure". According to the Theory of Relativity, pressure (or energy), like mass, causes attraction. Negative pressure causes repulsion.

Following the publication of the supernova discovery, additional evidence for the accelerating expansion of the universe began amassing. One of the main pieces of evidence was based on an analysis of the temperature fluctuation pattern of the cosmic background radiation measured by the WMAP satellite, as described previously. Another piece of evidence is based on measuring the redshift of 200,000 galaxies throughout the universe, and a comparison to models of the formation of structures in the universe. It is important to mention that the various evidence that has accumulated (and continues to accumulate), which is based on different measurement methods and analyses of various data (the cosmic background radiation, galaxies, supernovae, and so on), provide a consistent picture of the behavior of the universe. The assemblage of data leads to the conclusion that the dark energy, which causes the universe to accelerate, constitutes some 73% of the energy content of the universe.

But what is that dark energy? We do not yet have an answer to that question. Indeed, it would be true to say that researchers today still have no clue regarding the nature of this strange energy. Several general ideas exist. According to one school of thought, dark energy reflects the *vacuum energy* of the spacetime, which is sometimes also called the *cosmological constant*. The concept of vacuum energy is taken from quantum field theories, according to which even in a space that is completely devoid of matter, the uncertainty principle necessitates a kind of quantum vibrations that carry energy. The nature of the vacuum energy depends on the specific model, which is still unknown. In any case, this idea has some intrinsic problems. Another school of thought purports that dark energy is related to a kind of basic particle called *quintessence*, which develops with the universe. As it is today, there is no evidence of the existence of such a particle and it is, of course, possible that still another explanation exists, which nobody has thought of yet, and is waiting for the next Einstein to arrive.

8. The Inflationary Stage — A Cosmic Booster?

Despite the phenomenal successes of the Big Bang theory and the support given by the observations, not all was cheery. Some weighty questions remained unanswered. The nature of these questions was essentially philosophical, and they focused on the fact that to create the universe in which we live, such unbelievable coincidences would have had to occur, that the probability of them actually occurring is infinitesimally small. These problems could have been solved using the *anthropic principle*, which purports that what seems to us to be an impossible coincidence is an obvious result of the requirement that the evolution of the universe enables the existence of intelligent life — because life in the universe is a fact. However, many in the scientific community regarded this as no more than a sophisticated way of saying that "this is the way things are, and had they not been like this, they would have been different", instead of seeking a profound physical explanation.

The Big Bang theory suffered from two main problems:

(1) *The horizon problem* (also referred to as the homogeneity problem): As mentioned earlier, the universe is expanding according to Hubble's law, in other words, the relative velocity of two galaxies increases in direct proportion to the distance between them. This velocity stems from the expansion of space itself, and so can exceed the speed of light. It therefore results that at a given time, distant regions in the universe are disconnected from each other. That is to say, a ray of light emitted by a given galaxy can travel only a finite distance during the time from the Big Bang until the present time. And since the speed of light has the highest velocity in the universe, this is true for every piece of information transmitted by the galaxy. The immediate consequence of this is that, at any given time, any observer in the universe can see only a finite region around him or herself. Light rays emitted from more distant regions have not yet reached him or her, and so he or she cannot communicate with those regions. The surface surrounding this region (see Figure 6) is called the *particle horizon* (as opposed to the event horizon of a black hole, which will be discussed later), and is the farthest surface the observer can see (hence the name "horizon"). As the figure illustrates, at any given time the universe is divided into regions that are unable to communicate with each other. Communication is possible only between points that are within the horizon. As time passes, the particle horizon grows since increasingly distant rays of light can reach that same point. The *visible universe* is the region of the universe which we, on Earth, can see today, with a radius of

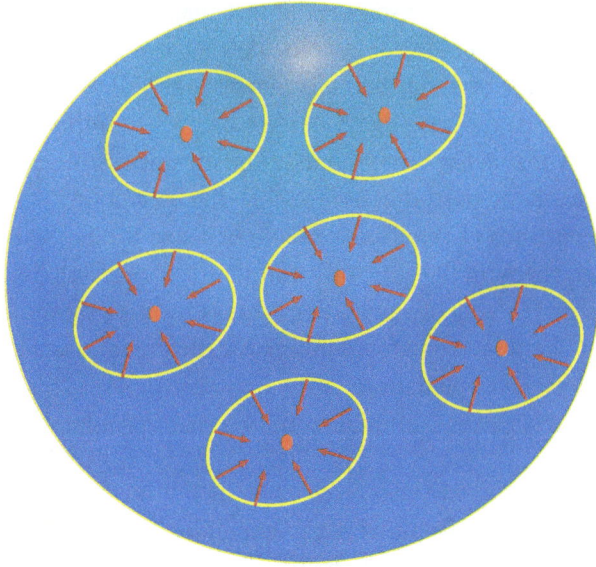

Figure 6. Each of the yellow circles denotes the particle horizon of its central orange point at a given time. Rays of light (red arrows) that leave the inside of the region encircled by a yellow circle, will reach the central orange point, making the source that emitted them visible. Rays that are emitted from points outside this region will not make it to the orange observer. The yellow horizon is the most distant surface the observer can see at this time. As time passes, the yellow circles around each of the observers grow.

approximately 46 billion light years. This datum seemingly contradicts the age of the universe — 13.7 billion years — since if the speed of light is the highest in the universe, how can the size of the visible universe be almost three times the distance that light can travel in a time that is equal to the age of the universe? The answer is that the light we see indeed did not travel more than 13.7 billion light years; however, during the time in which the light traveled from its source to Earth, the universe continued to expand, and the distance between the source and Earth continued to increase.

During the age of recombination, the particle horizon was smaller than it is now. Its angular size, as it would have appeared to a present-day observer, is about four times the size of the Moon. Hence, the present-day visible universe contains thousands of regions (yellow circles in the above figure) that during the age of recombination were causally disconnected from one another, and the question arises: Why are the temperature fluctuations detected in the microwave background radiation so small? If the temperature and density of a certain region had changed before the age of recombination, other regions would not have been able to respond to that change, since the information would not have reached them in time. After the age of recombination, radiation had already separated from matter, and there was nothing to affect it. It seems as if something "froze" the state of matter immediately after the universe was created until the age of recombination. This is referred to as the horizon problem.

(2) *The flatness problem:* As we have seen, Friedmann's equations offer various solutions depending on the density of matter and energy in the universe. There is a special density value, called the *critical density*, for which the universe is flat. Observations reveal that of all of the infinite possible values, the density in the universe is indeed the critical density as far as the measurement accuracy is concerned (in other words, the universe is flat). Even a minute deviation from this value in the first split second after the creation of the universe would have caused it either to collapse at a very early age, before the galaxies were formed, or to expand too rapidly. The fact that we exist indicates that something set the density to the exact correct value, yielding a flat universe, so that we could be created, as if it were the hand of God. This is called the flatness problem.

An elegant solution to these problems arrived in 1980 in the form of the *inflationary model* proposed in parallel by Alan Guth and Alexei Starobinsky. According to this model, the universe underwent a kind of rapid inflation that took place from approximately 10^{-35} secs after the Big Bang to about 10^{-32} secs after it. In this fleetingly short time, the universe inflated from an extremely small size — 10^{-33} cm — to a size of about 10 centimeters. This rapid inflation rate, which is millions of times faster than the speed of light, froze the initial quantum fluctuations that were created before the inflation stage, while the universe was still in causal contact. In other words, the universe blew up so quickly that there was no time for substantial inhomogeneity to develop. This "smearing" of the disturbances is what caused the uniformity of the background radiation. The rapid inflation similarly caused the universe to be flat. To understand this, imagine that we are living on the surface of a sphere (a closed universe) with a radius of 10 m. If we walk a distance of several meters on this spherical shell, we will undoubtedly notice its curvature. Let us now suppose that the ball has suddenly inflated and is the size of the Earth. If we walk a distance of several meters now, the expanse will seem flat to us (indeed, in our daily life we do not notice the curvature of Earth). Since the visible universe is limited in size, as explained above, the inflation phase caused the part of the universe we can measure to seem flat.

What caused the universe to inflate so rapidly? According to one paradigm, it was a hypothetical particle, called an "inflaton", whose properties resemble those of the "Higgs particle", which was discovered by scientists at the LHC collider in 2012. Like dark energy, this ancient particle created a kind of anti-gravity force, which caused the universe to inflate at an increased rate. At the end of the inflationary stage, this particle's energy turned into the elementary particles we are familiar with, and from there on the universe continued to evolve according to the original Big Bang theory.

Nowadays, the inflationary stage is part of the cosmological model. Since this stage was first proposed, a strong connection has been revealed between the elementary particle theories and the evolution of the universe. This perception led to the development of a new branch in physics called *particle cosmology*, which attempts to delve into the depth of this relationship. Nevertheless, it should be noted that there is more to this than meets the eye, and that many believe that the latest discoveries, both in cosmology and in particle physics, indicate that an essential component of the theory is still missing.

Chapter 2

The Genesis of Galaxies
and the Birth of Giant Black Holes

1. Galaxies — The Natural Habitats of Stars

As explained in Chapter 1, the distribution of matter in the universe during the initial stages after its formation was almost completely uniform, with the exception of very small primordial fluctuations, of a magnitude of about 1:100,000, as the COBE and WMAP satellites discovered in the cosmic radiation pattern. At some stage after the universe had cooled down sufficiently, these fluctuations caused slightly denser-than-average regions to collapse and compress due to their self-gravitational force, and create large masses of dark matter and gas. At a certain time, stars began to form within these dense masses of matter, and eventually turned into the galaxies and other structure that are visible in the universe. According to various estimations, the first structure began to form when the universe was still very young, less than 100 million years old (about one hundredth of its current age). Direct evidence of this is provided by in-depth observations conducted using the space telescope and ground-based telescopes, and especially the giant Keck telescope in Hawaii, which revealed galaxies that were formed 13.2 billion light years ago, namely, when the universe was less than 500 million years old.

A galaxy is a collection of stars that are held together by their mutual forces of attraction, pervaded by tenuous matter that consists of gas and dust particles. The visible universe contains some 200 billion galaxies of different and varied shapes, which are commonly divided into three main types: spiral galaxies, elliptical galaxies, and irregular galaxies (with no defined shape).

Spiral galaxies are flat and disc-shaped with spiral arms (hence their name) and a galactic bulge in the center. The number of spiral arms and their shape varies from galaxy to galaxy, and both the arms and the galactic bulge are clearly visible when viewed face-on. The gas and stars in the disc revolve around the center of the galaxy with a speed that increases with their distance from the galactic center, as revealed by measuring the spectrum of the stars and using the Doppler Effect. The galactic spiral arms are regions of higher-than-average gas and dust density, and therefore most of the stars are created there, as attested to by their blue color. The blue light is emitted by young stars, which are warmer than older stars such as the Sun. The arms also contain concentrations of dense gas and dust clouds, creating conditions consistent with those expected in stellar nurseries (or star-forming regions), bearing evidence of stars in the process of creation. The galactic arms are, in fact, density waves traveling in the galactic plane at a velocity that differs from that of the stars in the disc, so that the stars and gas actually pass through the moving arms. As shown in Figure 1, the core area is redder than the arms, since it contains mainly older stars of lower temperature than the younger stars in the arms. Spiral galaxies contain between one billion stars in the smaller galaxies and hundreds of billions in the larger ones, and their diameters range from 30,000 to 100,000 light years.

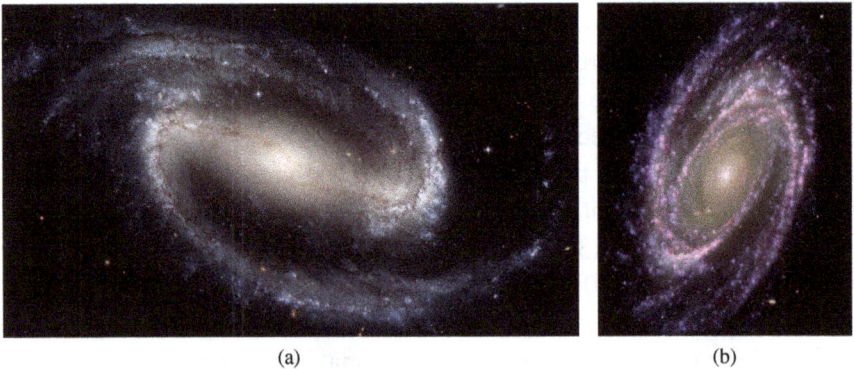

(a)   (b)

Figure 1.   Spiral galaxies viewed from above: M81 (b) and NGC 1300 (a). Most of the stars are formed in the spiral arms (imparting a bluish tint), where density is highest. (a): Courtesy of NASA, ESA, and A. Zezas. (b): Courtesy of NASA, ESA, and A. Zezas/ Galex/Spitzer.

(a) (b)

Figure 2.   NGC 4013 (a) and the Sombrero Galaxy, also called M104 (b), as viewed from the side. The galaxy was named "Sombrero" due to its resemblance to the broad-brimmed Mexican hat. The brown hue of the disc results from the absorption of star light by interstellar dust, which heats the dust, causing it in turn to emit infrared radiation. Courtesy of NASA and The Nicmos Group (STScI/ESA).

(a) (b)

Figure 3.   Infrared images of the Sombrero Galaxy (b) and M63 (a). The red disc seen in each of the images originates from the emission of infrared radiation by dust rings in the galactic disc, which are heated by the absorption of visible light and ultraviolet radiation emitted by the stars. The images were taken by the Spitzer Space Telescope, a satellite-mounted infrared telescope. (a): Courtesy NASA/JPL-Caltech/SINGS Team. (b): Courtesy of NASA/JPL-Caltech/R. Kennicutt.

Viewed edge-on, a large halo is visible above the galactic disc. This halo consists of globular clusters with old stars scattered amongst them. Globular clusters, which can be seen clearly even through relatively small telescopes, contain hundreds of thousands of old stars that are held together

(a)   (b)

Figure 4. Images of elliptical galaxies: NGC 2787 (b); NGC 1132 (a). Courtesy of NASA, ESA and The Hubble Heritage Team (STScI/AURA).

by mutual gravitational forces. These clusters were apparently created a very short while after the galaxy was formed, or even during its formation.

Elliptical galaxies are ellipsoidal[1] and come in a wide range of shapes and sizes (as shown in Figure 4). Their projection on the plane of the sky (in other words, their apparent, two-dimensional shape) is an ellipse with various degrees of oblateness, from a practically perfect circle to highly elongated ellipses. They range in size between 6,000 and 300,000 light years, and they contain between 100 million stars in the smallest dwarf galaxies and 200 billion stars in the giant galaxies. Elliptical galaxies are yellowish-orange since they contain mainly old stars. Unlike spiral galaxies, the stellar population in the center of the galaxy does not differ from that in the external parts of the galaxy. In fact, elliptical galaxies resemble the galactic bulge of spiral galaxies in both shape and color. It is commonly believed that the process of star formation in these galaxies has ended a long time ago. The movement of stars in elliptical galaxies is random and along radial directions, that is, from the center of the galaxy outward, so that these galaxies have almost no rotation. In this respect,

---

[1] An ellipsoid is a three-dimensional body created from the revolution of an ellipse around its axis.

Figure 5. Merging galaxies, as imaged by the Hubble Space Telescope. Courtesy of STScI/AURA-A.

they differ from spiral galaxies in which, as mentioned above, the stars revolve around the center of the galaxy.

The formation process of elliptical galaxies is not entirely clear. According to one explanation, elliptical galaxies were created as a result of the merging of spiral galaxies. Such collisions were relatively frequent when the universe was younger. Merging galaxies were observed frequently, and many believe that this is also the origin of the *irregular galaxies* (as shown in Figure 5). The elliptical galaxies, unlike the irregular galaxies, apparently had enough time to virialize and attain their shape.

2. Giant Black Holes in the Centers of Galaxies

Another kind of black hole exists, whose mass is significantly greater than those black holes that are formed after the death of stars. These black holes are called *supermassive* or simply *giant black holes*, and their masses range between one million and one billion times the mass of the Sun. The largest black holes ever discovered are the size of the inner part of our solar system — approximately one billion kilometers in size. The existence of supermassive black holes was first deduced in the 1960s when quasars were first discovered. Recent research reveals that in fact every galaxy, including our Milky Way, hosts a supermassive black hole at its center. In some galaxies, this black hole is active, accreting large quantities of matter and emitting strong radiation and jets of matter, but in most galaxies it is quiet. Galaxies with active black holes are called *active galaxies*. The over luminous region at the center is called an *active galactic nucleus*. Quasars, which will be dealt with later on, are examples of such galaxies, in which the luminosity of the radiation emitted from the vicinity of the black hole exceeds 10-fold that of the entire galaxy.

How are giant black holes created? The answer to that question is not entirely clear. The current paradigm is that they started as small black hole seeds at the center of young galaxies, and then grew rapidly through accretion of matter transferred to their vicinity from the outer parts of the galaxy by some violent processes, such as galaxy mergers. There is accumulating evidence that supermassive black holes with masses in excess of 10 billion solar mass were already present at early cosmic epochs, when the universe was merely one billion years old (less than one tenth of its present age), and that at that epoch they accreted matter at a relatively large rate. Whether they had enough time to reach their final masses at such an early cosmic time is an issue currently under debate in the astrophysical community. Some claim that supermassive black holes can only reach such huge masses provided the seeds out of which they grew were massive enough, at least 10,000 times more massive than ordinary black holes that are produced in stellar explosions. What is the nature of those black hole seeds then? One idea is that under certain conditions, giant stars may have formed in the early universe, with masses that are 10,000 and more times that of the Sun. These stars were unstable and collapsed directly into a black hole within a very short time, without creating a supernova. According to yet another model, giant black hole seeds resulted from the collapse of large and dense clusters of stars. Others claim that the rate at which a black hole can, in principle, absorb mass in its early stages is so high that even seeds made of ordinary black holes may suffice. Whichever the correct theory, most researchers agree that the existence of supermassive black holes follows from the observations. The best evidence for the existence of supermassive black holes is found in the center of our galaxy, as we shall presently see.

3. The Milky Way Galaxy and the Black Hole in its Center

The Milky Way Galaxy, which is home to our solar system, is a typical spiral galaxy, similar to those presented in the above figures. The diameter of the galactic disc is 100,000 light years and it is only 1,000 light years thick, in other words about one hundredth of its diameter. It contains approximately 200 billion stars. The number of the spiral arms and their exact shape are not known for certain; four arms have been identified, but there are possibly additional arms that are not visible to us due to our location in the galaxy. Contrary to what many people believe, our solar system is not at the center of the galaxy, but is located some 25,000 light years from the center (see Figure 6). The tail of white light we see when we look up at the sky on a dark night is the cumulated light from billions of stars that make up the galactic disc (hence the name Milky Way). The solar

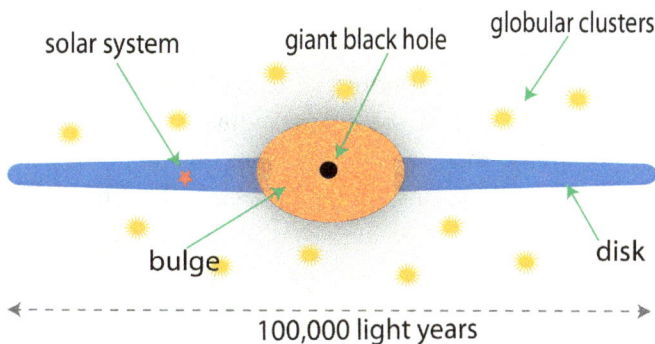

Figure 6. A schematic description of the Milky Way Galaxy as viewed edge-on.

system passes through the main arms about once in 150 million years. The center of the galaxy is in the direction of the Sagittarius constellation, where a giant black hole, whose mass is four million times the mass of the Sun, is located.

An especially luminous radio source was recently discovered in the Sagittarius constellation, adjacent to the estimated location of the black hole. Some believe that this source results from activity related to the black hole. Bursts of radiation of an unknown nature have also been detected in recent years in the black hole region. According to one of the more common interpretations, these eruptions are caused when matter that accumulates around the black hole suddenly falls into it, heats up, and emits radiation. Despite this evidence, the black hole is at present quiet. Even if the eruptions originate in the black hole, as many researchers believe, they are relatively weak and so are not indicative of activity such as that seen in quasars and active galaxies, in which, as we will describe in detail in Chapter 16, the black hole absorbs tremendous amounts of matter at a monumental rate. Evidence obtained lately reveals that the black hole located at the center of the Milky Way was possibly active in the past, several thousands of years ago.

The strongest evidence for the existence of a giant black hole at the center of the Milky Way comes from measuring the orbital motions of stars around the center of the galaxy. Black holes are the densest objects in the universe. In order to know for sure whether something is indeed a black hole, its density must be measured. Density cannot be measured directly, but it can be deduced by measuring the mass and volume of the body. Using special techniques developed for the viewing of infrared radiation, astronomers succeeded in measuring the location of stars at the center of the galaxy with unprecedented accuracy. An analysis of the motions of a large number of stars enabled researchers to identify the position of the massive object that attracts them. One of the stars, denoted S2, is located very close to the position of that unidentified object and scientists decided to measure its orbit with high precision. Since the orbital time of S2 around the center of the galaxy exceeds 20 years, prolonged measurements were needed. From 1992 to this date, German and American astronomers have been observing the

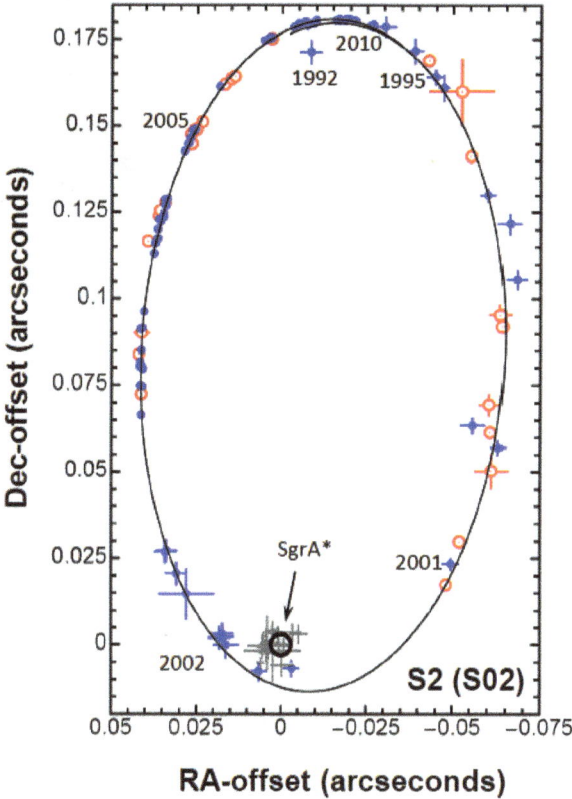

Figure 7. The ellipse represents the motion of S2 around the black hole located at the center of the Milky Way (center of the small circle). Each of the crosses represents a measurement of the star's location. The year in which the measurement was taken appears aside each cross (data presented were measured between 1992 and 2010). Courtesy of Reinhard Genzel. Reproduced with permission of the AAS.

movement of this star using specially developed telescopes (see Figure 7). An analysis of the star's trajectory enables scientists to deduce the mass of the central object and estimate its radius and density. No known object, other than a black hole, can account for the measured orbit of the S2 star.

Chapter 3

Stellar Evolution — From Dust Cloud to Black Hole

Like all living creatures, stars are born, they live, and they die. They are created within dense dust and gas clouds like that presented in Figure 1. This happens when a certain region of a cloud is slightly compressed and cooled. As a result of the gravitational force acting on it, that region becomes even more compressed and a *protostar*, which will later become a star, is formed. This stage is relatively short and lasts only several million years. When the density and temperature at the center of the protostar reach sufficiently high values, nuclear reactions at the center of the protostar "ignite", and a process of nuclear fusion occurs, like the process that takes place in a nuclear fusion reactor, turning hydrogen into helium. The instant at which the fusion reaction ignites marks the star's moment of birth. A prodigious amount of energy is released during the nuclear fusion process, most of which turns into heat and radiation.

The great pressure applied within the star by the hot gas balances the gravitational force, and hydrostatic (hydro = liquid, static = rest) equilibrium is achieved — in other words, a steady state in which the star is at rest and its radius is constant. The temperature of the star is not uniform: the temperature at its center reaches tens of millions of kelvins, while on its surface it is only several thousand kelvins. The temperature on the surface of the star determines its color: hotter stars appear bluer, whereas colder stars are red. The luminosity of a star and its internal temperature (or alternately its color) maintain a simple relationship: the brighter the star, the higher the temperature on its surface, and the bluer its color. This

Figure 1. A gas cloud system in which new stars are created. Image taken by the Hubble Space Telescope. Courtesy NASA, ESA, STScI, J. Hester.

relationship is called "main sequence", and stars that maintain it (in fact, all ordinary stars) are called main sequence stars (see Figure 2).

This relationship depends mainly on the mass of the star: the heavier the star, the brighter and bluer it is. The age of the star, which is defined as the length of time the nuclear core has been active, also depends on the mass of the star created: the greater the mass, the shorter the lifetime. The lifetime of our Sun is approximately 10 billion years. A star with a mass 10 times that of the Sun will live only one thousandth of that time, in other words, only 10 million years. This is due to the fact that the greater the mass of the star, the higher the combustion rate of the nuclear core, causing

Box 1.   Nuclear Fission and Nuclear Fusion

Nuclear Fission is a process in which the nucleus of a heavy atom, such as Uranium 235, is split into two lighter nuclei, releasing energy in the process. Energy is released because the mass of the heavy nucleus is greater than the sum of the masses of the fission products (in other words, the bond energy of the fission products is greater than that of the original nucleus). This difference in masses turns into energy, which is released in the form of radioactive radiation, as required by the special theory of relativity. Fission takes place in atom bombs and nuclear fission reactors.

Nuclear Fusion is the opposite process of nuclear fission: two light nuclei merge into a heavier nucleus with the release of energy. In this case, energy is released due to the fact that the mass of the fusion product (the heavy nucleus) is smaller than the sum of the masses of the two light nuclei that underwent fusion. Fusion takes place in hydrogen bombs, fusion reactors, and the cores of stars, in which four hydrogen nuclei fuse to become one helium nucleus.

Iron is the most stable element in nature (in other words, the iron nucleus has the highest binding energy). For nuclei that are lighter than iron, the binding energy is smaller. This trend is reversed in nuclei that are heavier than iron. Hence, fusion is possible only for nuclei that are lighter than iron, whereas fission takes place only in nuclei that are heavier than iron.

hydrogen, which is the nuclear core's fuel, to deplete at a faster rate. This is also the origin of the main sequence temperature–luminosity relation described above, which often serves as an observational means of determining the age and mass of a distant star. The destiny of the star, after all of the hydrogen in its core is depleted (about one tenth of the star's mass), depends on its initial mass. If this mass is much smaller than that of our Sun, the core temperature will be lower than that needed to activate helium fusion. In such a case, the star will slowly cool down and shrink, and will appear as a small reddish star called a *red dwarf*.

The evolution of stars whose mass is equal to or greater than that of the Sun is different. These stars are the factories in which the densest objects in the universe are manufactured, as described in Figure 3. In such stars, the core begins to collapse due to the monumental gravitational

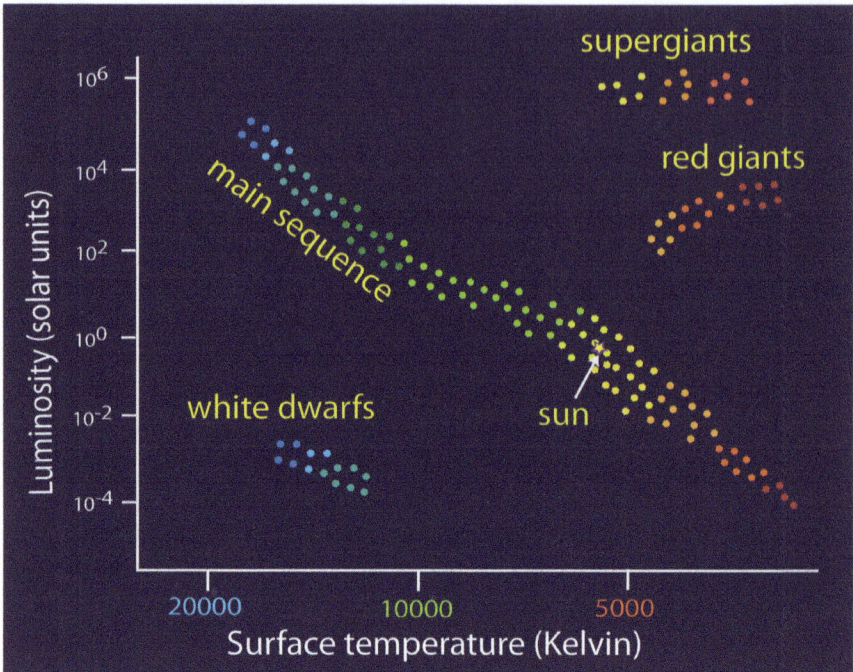

Figure 2.   The relationship between a star's luminosity and the temperature on its surface (or alternately, its color). Each point represents a measurement of a star in the galaxy. Ordinary stars, like our Sun, in which the nuclear reactor is active, are called main sequence stars, and they maintain a simple relationship (the large group of points in the middle), which depends on the mass of the star. The blue points at the top left-hand side of the main sequence are the heaviest stars, while the red points at the lower right-hand section are the lightest. The post main sequence stars (supergiants, red giants and white dwarfs) are also indicated in the diagram.

force acting there. This collapse, in turn, causes the layer of matter above the core, which mostly contains hydrogen, to heat up, triggering the fusion of hydrogen to helium in this layer. The energy released in the layer causes the star to be 1,000–10,000 times brighter than it was originally. The external layers of the star inflate to fantastic dimensions due to the monumental pressure suddenly created, and the original star becomes a red giant, which is a kind of cold star that is 100 times, or more, larger than the original star. In parallel to this process, the core continues to shrink and heat up, until at some point, when the core temperature reaches

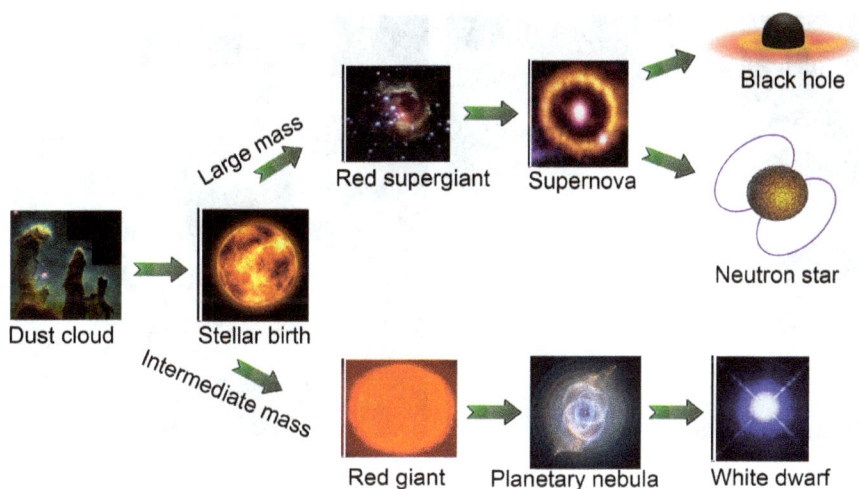

Figure 3. The different stages a star goes through from birth to death, at which point it becomes a compact object — a white dwarf, a neutron star, or a black hole.

100 million kelvins, helium nuclei accumulated in the core begin to fuse — a process in which three helium atoms become one carbon atom.

In medium-mass stars (with masses that do not exceed 10 times the mass of the Sun), the core temperature is not high enough to begin the fusion of carbon produced from the combustion of helium. In this case, the core will become a white dwarf — a star in which the gravitational force is balanced by the quantum pressure of electrons — while the external layers of the red giant are expelled by strong stellar winds to become a strikingly beautiful planetary nebula, such as those seen in Figure 4. This is also the future that awaits our Sun, which will become a red giant in a few billion years. Astronomer William Herschel coined the name *planetary nebula* to refer to these gas clouds, which reminded him of the appearance of the planet Uranus, although planets and planetary nebulae are completely unrelated. In the center of each nebula is the remaining core remnant of the collapsing star in transition to becoming a white dwarf. The nebula itself contains ionized gas, mainly hydrogen and helium, as well as small amounts of additional elements. Strong ultraviolet radiation emitted by the hot star in the center causes the nebula to heat up and emit the observed radiation.

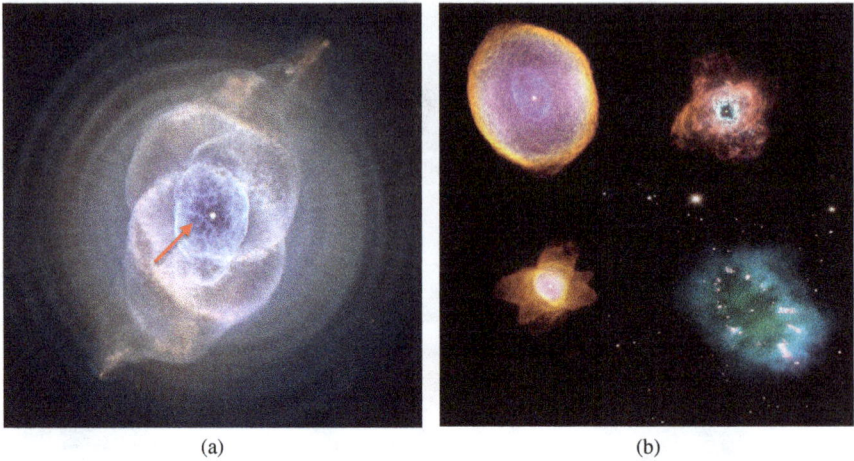

Figure 4. (b) A mosaic of planetary nebulae. A white dwarf is located at the center of each nebula. (a) The Cat's Eye Nebula and the white dwarf at its center (red arrow). Courtesy NASA, ESA, HEIC, and the Hubble Heritage Team (STScI/AURA).

Box 2.   Why are Planets not Stars?

An ordinary star, like the Sun, is a gaseous mass with an internal energy source. The source of energy is the nuclear reactor that operates in the core of the star. Planets, such as Earth, Mars, Jupiter and Saturn, do not have a self-sustaining source of energy that causes them to radiate. Their energy supply comes from the Sun, or in the case of extra-solar planets, from the mother star around which they revolve. So while the light emitted by the Sun is the product of the Sun being heated by the nuclear core in its center, the light emitted by the planets is in fact light from the Sun that reached the planet and was reflected back. In order for the nuclear core to operate, the mass of the star must exceed 8% of the Sun's mass. Objects with smaller masses have no internal energy source. This group of bodies includes the planets, and another class of objects called brown dwarfs.

The magnificent colors of the nebulae originate in the atomic emission lines of the various elements that make up the nebula. The lifetime of a typical nebula is about 10,000 years, a very short time compared with the lifetime of a star. After this time, the central star cools off, stops emitting ultraviolet radiation, and becomes a white dwarf. As a result, heating of the nebula ceases, and it stops emitting radiation. Some 1,500 planetary nebulae have been observed in our galaxy.

In stars with a mass more than 10 times that of the Sun, fusion of carbon will begin, followed by nuclear reactions of heavier elements, until iron, which is the most stable element in the periodic table. After that, no more fuel will remain in the core to balance gravity, and it will collapse into a black hole or a neutron star within several seconds. This collapse will cause the external layers to be sent off at high velocities and form a mighty blast wave that emits radiation. This will seem to spectators on Earth like an explosion called a *supernova*. The story of these explosions will be told later on.

Second Episode

Physics and Astronomy in the 21st Century

The mechanisms responsible for the plethora of exotic phenomena that will be discussed in the third part of this book are based on principles of modern physics; the ways in which these phenomena are observed are based on methods of modern astronomy. This episode will begin with a review of the nature of the matter and forces that exist in nature, and will continue with a description of the various types of radiation that objects in the universe emit, and the methods we use to measure them.

Chapter 4

Matter, Forces, and Symmetry in Nature

1. The Structure of Matter

Our everyday experience tells us that all matter around us is divisible. It can be broken, melted down, even crumbled into a thin powder. Can any kind of material be broken down into components that are as small as we desire, or is all matter composed of very small building blocks that are themselves indivisible?

Human beings have been preoccupied with the structure of matter since the beginning of time. Greek philosophers in ancient times developed several theories that aimed to address this issue. One of the more accepted theories purported that all matter in the world is composed of four elements: air, water, earth, and fire. It is interesting to note the distinction, according to which each of these elements represents one of the states of matter: gas, liquid, and solid. And fire? Fire represents radiation. The Four Elements Theory prevailed until the Middle Ages but was abandoned in the modern age. According to another theory developed in ancient Greece by Greek philosopher Democritus, matter is made up of tiny particles that cannot be further divided, which he called atoms (*atomos* means indivisible in Greek). According to Democritus' theory, the universe consists of a vacuum and atoms that connect to similar atoms and create the stars and matter that are in the universe. Different materials are made up of different kinds of atoms and can change their shape by changing the composition of the atoms.

The atomic model of matter was supported by the pioneering works of English chemist John Dalton in the early 19th century. These works served as the basis for the construction of the periodic table in 1869 by

Dmitri Mendeleev, in which the different elements are classified according to the atomic number and chemical symbol of the atom they are made up of. The physical explanation of the relationship Mendeleev discovered was provided only several decades later, after the development of the quantum theory. Research conducted at the end of the 19th century and the beginning of the 20th century revealed the components of the atom itself, and it soon became clear that the atom consists of a nucleus that is surrounded by electrons. The rapid development of particle physics, a branch of physics that began developing in the mid-20th century, led to the understanding that atoms are not the basic building blocks, but rather are subject to additional division themselves. These studies led to the discovery of an entire menagerie of particles, called elementary particles, and to the classification of the various particles discovered, based on symmetry properties. Despite thousands of years of research and investigation into the nature of matter, we still have no definite answer to the question: what is the most elementary particle in nature?

The commonly accepted picture today is that any material — water, air, wood, iron, and so on — is made up of a substructure, as illustrated in Figure 1.

The basic units that determine the properties of matter are called *molecules*. Each individual molecule is a cluster of several (identical or different) atoms. The number of atoms in a molecule varies from material to material. Water, for example, contains molecules made up of only three atoms each: one oxygen atom and two hydrogen atoms. Organic substances, on the other hand, contain giant molecules that are made up of dozens of atoms each. Every individual atom has a nucleus with a positive electrical charge, which is surrounded by electrons, each of which has a

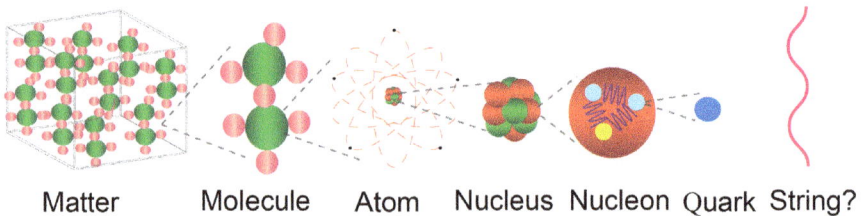

Matter   Molecule   Atom   Nucleus   Nucleon   Quark   String?

Figure 1.   The structure of matter.

negative electrical charge. The electrical attraction between the negative electrons and the positive nucleus binds the electrons to the nucleus and prevents the atom from falling apart. The nucleus itself contains protons and neutrons, which together are called *nucleons* (components of the nucleus). The number of protons in the nucleus — which is also called the *atomic number* — determines what kind of atom it is as well as its chemical properties. The neutrons, as will be explained later, constitute a kind of adhesive that prevents the nucleus from flying apart.

Immediately after the atomic structure was revealed (as shown in Figure 2), indications began to appear to the effect that the atom components may, in fact, not be the most elementary particles that exist. The first indication of this was the phenomenon of radioactivity, discovered in 1896 by French physicist Henri Bacquerel, who noticed a highly penetrative radiation emitted from uranium salts. Studies that were conducted in the two decades following Bacquerel's discovery revealed that the radiation is emitted from unstable atoms and can be classified into three kinds: alpha,

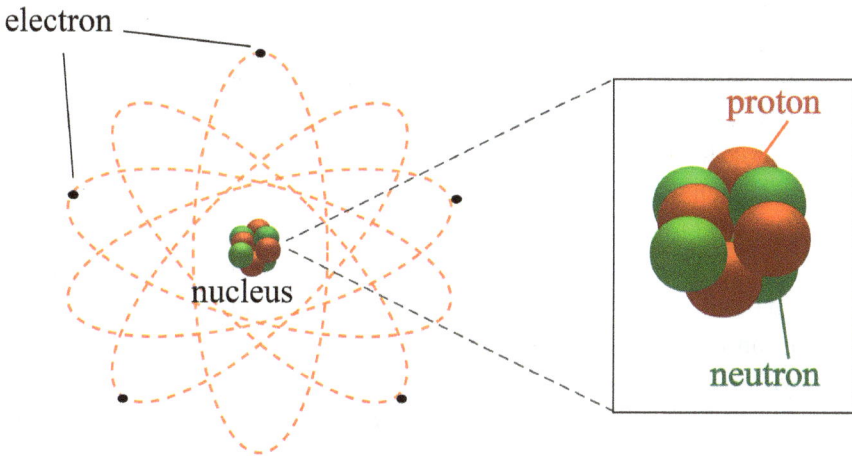

Figure 2. A schematic description of the structure of an atom: the atomic nucleus is made up of positively charged protons and neutral (chargeless) neutrons that constitute the adhesive holding the nucleus together. The number of protons in the nucleus determines what kind of atom it is (for instance, hydrogen has one proton, helium has two, and carbon has six). Negatively charged electrons surround the nucleus, revolving in fixed orbits. The average distance between the electrons and the nucleus is 100,000 times greater than the diameter of the nucleus itself.

beta, and gamma. The surprise came when beta radiation was actually found to consist of electrons that are emitted from an atomic nucleus, and that during the radioactive decay process, an atom that emits an electron becomes a different atom. The fact that an atom can suddenly turn into another atom led to the conclusion that the components of the nucleus are not elementary particles: they can change shape, and in order to do so they must be made up of even more elementary units, like a structure that is made of Lego® bricks and whose shape may be changed by changing the arrangement of the bricks.

The detailed mechanism responsible for radioactivity was fully understood only after the neutron was discovered in 1932 and the structure of the nucleus was revealed. Various experiments proved with certainty that the neutron is not an elementary particle, but that it tends to decay into a proton, an electron, and another very strange particle called a *neutrino*. Furthermore, scientists showed that during the radioactive decay process, neutrons in the nucleus of the original atom turn into protons, ultimately yielding a different, more stable atom. We will revisit the process of the neutron's decay and the emission of neutrinos later, when we discuss the role of this strange particle in astronomy.

The evidence that accumulated regarding the nature of the atomic nucleus led eventually to the discovery of the elementary particles it is made up of. It was found that all nucleons, meaning every proton or neutron, are each composed of three elementary particles called quarks, which are held together by a strong force that prevents them from separating from one another. The properties of quarks and of the forces that act between them will be described later in Chapter 12. For now, it will suffice to mention that there are, up to date, six known natural types of quarks, from which a huge variety of more complex particles may be constructed. Particles made of quarks are called *hadrons*. Protons and neutrons are the most well-known hadrons, but there are many other kinds of hadrons as well, which have been discovered in thousands of experiments conducted in various particle accelerators over the years.

Some particles that exist in nature are not hadrons, or in other words, are not made up of quarks. The best known of these is the electron, an indivisible elementary particle. Two additional, non-hadronic, negatively charged particles are *muons* and *tau particles*, which are identical to the electron in all properties except mass. Tau particles are the heaviest of the three, followed by muons, and then by electrons, which are the lightest. Neutrinos, which were mentioned earlier and of which there are three known types, are also elementary particles. These six elementary particles — the three charged particles and the three neutral particles — constitute a family, similar to the quarks. In order to distinguish them from the quark family, they were given the general name *leptons*. In summary, there are 12 known elementary particles of matter in nature, which are divided into two families of six members each — the quark family and the lepton family.

Although there are 12 elementary particles in nature, it turns out that the ordinary matter that surrounds us, which makes up the stars and galaxies, our tables and chairs, the air in our room, and even our human bodies, is composed of only two kinds of quarks and electrons. This tremendous variety is then, ultimately, a collection built from three basic building blocks; it is as if the entire universe is a kind of huge game of Lego®. The remaining quarks and leptons, which are not ordinarily present, can be created in large particle accelerators. These elementary particles are unstable and decay very shortly after they are created in the accelerator. They are also occasionally created in the atmosphere that surrounds the Earth, when a stream of energetic particles, called *cosmic rays*, penetrates the atmosphere from the outer space and collides with the air atoms. In this case too, the particles decay shortly after their creation, but they can nevertheless be detected before they decay. As we will explain later on, these particles play an important role in some of the cosmic ray detectors.

Are quarks the building blocks? According to the string theory, an even more basic entity exists in nature — the superstring — and the quarks themselves, like all other particles in nature, are in fact vibrations of this string. This may be likened to a guitar string; its vibrations create sound waves that reach our ears and are heard as sounds. A variety of sounds may be produced from one single string by causing it to vibrate in different ways. Likewise, according to the superstring theory, quarks and other particles are the different sounds that the superstring produces. No experimental proof of the validity of this theory exists, and at present, it is still only a theoretical idea. Indeed, some scientists question it.

2. Quantum Theory and the State of Matter

All of the above still does not account for the properties of matter on the microscopic level. What enables atoms to exist over time? Why does nature boast a variety of chemical elements, and what shapes the periodic table of these elements? What is the origin of the force that resists us and prevents atoms from crowding together when we try to forcefully compress matter? Why do some materials conduct electricity while others act as insulators? The short answer to all of these questions is "quantum theory", and more specifically, two basic principles of this theory which we will now address: Heisenberg's uncertainty principle and Pauli's exclusion principle.

Before we begin reviewing the developments that led to the formulation of the quantum theory, we should mention that this theory is the strangest theory, the most difficult to grasp, and the theory that contradicts our intuition and daily experience more than any other theory in modern physics. One of the main reasons for this difficulty stems from the fact that quantum effects are "felt" on atomic and subatomic scales, but are indiscernible in daily life (unlike the gravitational and electromagnetic forces, for instance, which dictate most of the phenomena we experience in our surroundings). Any physics student first studying the foundations of quantum mechanics feels confused and perplexed in the face of the difficulty to understand the profound concepts underlying the fundamental equations of quantum mechanics; at the same time, however, she feels elated by the explanations of various natural phenomena that this theory provides. Even Einstein himself, who played a key role in developing this theory, later objected to some of its main ideas, which to others seemed inevitable (in this matter, it eventually turned out that he was wrong). Furthermore, although most researchers agree that the basic equations of quantum mechanics indeed describe nature, there are several interpretations as to the philosophical essence of the quantum theory. The quantum theory also contains some paradoxes — internal contradictions — some of which have already been given explanations (with which not everyone agrees), while others are still waiting to be resolved. Over the years, Einstein himself raised some of the more well-known and profound paradoxes, during his attempts to pinpoint logic failures in the ideas he opposed. It is ironic that the attempts to deal with paradoxes raised by Einstein (especially those made by Niels Bohr, who will be mentioned later), are those that contributed the most to the establishment of quantum mechanics.

The quantum theory began with an attempt to understand the characteristics of black-body radiation. A black body is the name given to matter that fully absorbs all electromagnetic radiation. Due to this property, such matter also emits radiation in a perfect manner. The characteristics of radiation emitted from a black body do not depend on the type of matter it is made of, the way in which it was heated, or any other detail, but solely on its temperature. When a black body absorbs radiation that falls upon it, it heats up, and as a result begins to emit radiation and cool back down. When a black body is at equilibrium, its temperature and the spectrum of the radiation it emits self-adjust so that its heating rate is equal to its cooling rate. Such ideal matter does not really exist, but many natural materials act, to a good approximation, as black bodies. At temperatures that prevail on Earth, black-body radiation is emitted in the infrared range, which is not visible to the human eye, and so such bodies will seem black to us (hence their name). The reason an asphalt road that is exposed to the sun heats up so much and seems so dark in color, is that it is, to a good approximation, a black body and so it absorbs the solar radiation that falls upon it in a very efficient manner. It also emits infrared radiation, which we cannot see. On the other hand, other objects that seem to be bright during the day are made of materials that do not absorb radiation in an ideal manner. When sunlight falls upon them, they absorb only part of the light and reflect (or disperse) the rest of it in all directions; this is the light that our eyes detect. When a black body is heated to a high enough temperature (over 3,000°C), it emits radiation in the visible spectrum. This is the source of the reddish color of white-hot iron and of filaments in light bulbs. Earth itself emits radiation like an approximate black body, as do the Sun and the stars. The cosmic background radiation emitted by the universe in its early days also possesses characteristics of black-body radiation.

The characteristics of black bodies were initially investigated by German physicist Gustav Robert Kirchhoff in the mid-19th century, and it was he who coined their name. In the decades that followed, scientists made many unsuccessful attempts to develop a mathematical model that would accurately describe the emission spectrum of a black body. It was clear that something fundamental was missing from the theory.

Toward the end of the 19th century, a German scientist named Max Planck, who was serving as professor of physics at Bonn University at the time, began taking an interest in the subject. The model he proposed to explain black-body radiation assumed that the radiation is emitted from matter by tiny oscillators, whose action resembles that of springs or chords. Each such oscillator vibrates at a single characteristic frequency, which is also the frequency of the wave it emits, and different oscillators have different frequencies. These oscillators can be likened to the strings of a piano; when you play a certain key, it causes a small hammer inside the piano to hit a corresponding string, making it vibrate. This in turn causes the air around the string to vibrate as well, creating a sound wave that propagates in all directions. The tone we hear is the sound wave that causes our ear drum to vibrate. Each of the piano's strings vibrates at one characteristic frequency that determines the frequency of the sound wave it creates (as well as the pitch of the note we hear). When several strings vibrate simultaneously, the wave that is emitted is composed of several frequencies (which we hear as a chord). The oscillators in matter are like piano strings, but instead of sound waves they emit electromagnetic waves.

Planck was not clear on the nature of oscillators in matter when he first proposed his model. This is not really surprising, considering the fact that when Planck was working on the development of his theory, the electron had not yet been discovered, and the structures of the atom and of matter were not yet known. Furthermore, Planck discovered that in order to obtain full agreement between the mathematical model he developed and the experimental data, it was not enough to assume the existence of oscillators in matter, but it was also necessary to assume that the energy of the oscillators is emitted in discrete packets, or quanta (a quantum means a discrete quantity), of a basic energy unit that depends on the frequency of that specific oscillator. The basic energy unit emitted by an oscillator of a given frequency v, is given by the product hv, in which the constant h, which later on was named *Planck's constant*, is a basic constant of nature.[1]

---

[1] For those interested, Planck's constant equals 6.626×10^{-34} Joule·sec.

The total energy emitted by an oscillator or by a collection of oscillators of identical frequencies, must equal a multiple of this basic energy quanta. In 1900, Planck first presented his revolutionary idea at a meeting of the Berlin Academy of Science. In 1918, Planck was awarded a Nobel Prize for his work.

Despite the model's success in explaining the measurements, Planck and his contemporaries believed that the quantization assumption, namely, the need to assume that matter emits corpuscles of light, is not related to the properties of light itself, but rather to the matter's emission mechanism, whose nature was expected to become clear when the structure of matter is properly understood and a more advanced model is developed. It soon became clear, however, that it was not a question of the structure of matter, but rather of the properties of light itself. It was Einstein who, in 1905, provided the complete explanation of Planck's findings in his historical article, which we mentioned earlier, for which he received a Nobel Prize.

3. Bundles of Light Quanta

According to Einstein's model, light may be regarded, despite its wave-like character, as being composed of massless particles called *photons*. The energy of a single photon exactly equals a Planck unit, $h\nu$. In other words, light comes in discrete packets (quanta) of energy, which are integer multiples of the energy of a single photon. Light emitted from a flashlight is made up of photons; radio waves that our cellular phone receives are also made up of photons, but of a much smaller frequency than that of light photons. This model of a light source may be likened to a machine gun that shoots bursts of bullets; a single photon is like a single bullet that is discharged from the machine gun, while the ray of light, which contains many photons, is equivalent to an entire burst of bullets. The frequency, ν, or equivalently the energy packet, $h\nu$, determines the color of the emitted wave. For example, a photon that corresponds to the color green has a larger energy packet than a photon that corresponds with the color red, just like a bullet discharged from a machine gun has much more energy than a bullet discharged from a small pistol. The number of photons the radiation source emits determines the light intensity — the more photons, the brighter the light source. According to this model, a source of green light, for instance, emits numerous identical photons with the same energy, which corresponds to a green wavelength. The higher the photon emission rate of the source, the brighter the source will appear (but its color will not change). A source of red light also emits identical photons, but their energy is lower than that of photons emitted from a green light source.

Einstein's model also provided an explanation for another phenomenon, the *photoelectric effect*, which was also deemed peculiar at the time. Experiments conducted in the late 19th century revealed that light whose frequency exceeds a certain threshold and that is shined on a metal surface causes the metal to emit electrons. It was later found that the number of electrons emitted depends on the intensity of the light, but that the energy of the electrons depends only on its wavelength (or color). For example, when visible light was shone on a metal surface, no electrons were emitted at all, even when the light intensity was increased significantly. On the other hand, when ultraviolet light was shone on the same metal surface, the electron emission was detected, even at relatively low radiation intensities, and the higher the radiation intensity was, more electrons were emitted. The wave theory of light, which was considered the pinnacle of physics at the end of the 19th century, failed to provide an explanation for this phenomenon, which was called the photoelectric effect, and it remained a mystery. The requirement of an energy threshold for the emission of electrons was particularly puzzling.

The solution to this problem arrived in the form of Einstein's quantum theory of light, as described above. Most solid materials we are familiar with from our everyday life are made up of an ordered system of atoms (in other words, a crystalline structure). The electrons in insulating materials, which do not conduct electricity, are bonded to the atoms and cannot move over considerable distances. Most of the electrons in metals, such as aluminum, iron, and copper, are also bonded to the atoms, but a small number of electrons (the conduction electrons) are free to move over large distances within the metal. This enables metals to be good conductors of heat and electricity. Although some of the electrons can move about freely in the metal, they cannot escape from it. Indeed, in order to remove a single electron from the metal, a threshold energy, called work function, is required. That is, we must give the electron we wish to remove energy that exceeds this critical value. Each kind of metal is characterized by a different threshold energy (or a different work function). When light is shone on a metal surface, some of the photons hit the free electrons. If the light being shined is of a high enough frequency, so that the energy of each photon in the light ray exceeds the metal's threshold energy, a free electron that is hit by a photon may escape and be emitted by the metal, as observed in the experiment. The higher the light intensity, the more photons hit the metal, causing the emission of more electrons. Light whose frequency is lower than the threshold frequency will have no effect, regardless of its intensity, since none of the photons has enough energy to detach an electron. The description according to which a photon hits an electron and causes its emission from the metal is a particle-based description that requires Einstein's quantization assumption. No mechanism in the wave-based description of light can explain why electrons are detached from metals in this way.

The photoelectric effect has many applications in our daily lives; one common example is photoelectric cells used in command and control systems as well as in many other devices. We will revisit this application later in the context of light detectors in Chapter 6.

One of the more important conclusions from Einstein's work is that light has a kind of dual nature: sometimes it acts like a wave and other times like a particle, depending on the kind of experiment being conducted. This property is called the *wave–particle duality*. Refraction, reflection, and diffraction, which are responsible for a variety of phenomena such as rainbows, mirror image, the distortion of an image in a glass of water, the action of eyeglass lenses, and more, are related to the wave nature of light. On the other hand, phenomena such as the photoelectric effect, the emission spectrum of a black body, and scattering and absorption of light by atoms and molecules are related to the particle nature of light. Do not feel discouraged if the properties of light as they appear in Einstein's model seem strange and incomprehensible. Many scientists in Einstein's day, some of whom were top physicists, objected to this revolutionary idea and argued that it will eventually become clear that there is in fact no need for it.[2] The decisive proof for the existence of light particles — photons — was finally provided by Arthur Compton, after whom the *Compton effect* was named, and for which he was awarded the Nobel Prize in 1927. As we will see now, Einstein was not only correct, but it soon became clear that nature is even stranger than previously believed.

4. Schizophrenia and Uncertainty in the Quantum World

The atomic model of matter began to strike root following the publication of John Dalton's work in 1805, but at that time the nature of atoms was still largely unclear. Tables of atomic weights enabled scientists to estimate the mass of a single atom (the hydrogen atom of our days), but not its structure. A breakthrough in the understanding of the atomic structure occurred only in 1897, when British physicist Joseph John Thomson (known better as J. J. Thomson) discovered the electron through a series of experiments he conducted with cathode tubes, for which he was awarded the Nobel Prize in 1906. Thomson showed that the new particle he discovered has a negative charge and a mass that is at least 1,000 times smaller than that of the hydrogen atom. Following this discovery, Thomson also proposed a kind of atomic model, called the plum pudding model, in which the electrons are stuck within the atoms (like raisins in a raisin cake). Thomson's model was eventually disproved by subsequent experiments (in particular, the Geiger–Mardsen gold foil experiment conducted under the supervision of Errnest Rutherford), but one thing was agreed upon by all: the electron is indeed a particle. Well, actually everyone except for one student by the name of Louis de Broglie.

Following Einstein's work, De Broglie, a member of the French aristocracy and the seventh Duke of Broglie, proposed that the concept of wave–particle duality be applied to electrons as well as to other physical systems. If light waves can behave at times as particles, why then can't the electron, or any other particle, sometimes behave as a wave as well? He also

---

[2] One of the greatest sceptics was Robert Millikan, who spent 10 years of his life trying, in a series of experiments, to prove Einstein wrong. Millikan's ultimate conclusion was that Einstein was, in fact, right. In 1923, Millikan was awarded the Nobel Prize for his accurate measurements that (contrary to his original intentions) supported the quantization postulate.

defined the relationship between the mass of a particle and its wavelength (which was later termed the *de Broglie wavelength*). The problem with this suggestion is that our everyday experience teaches us that a particle can be in only one place at any given time, and that the trajectory of a particle is deterministic. In other words, if we throw a particle at a given velocity in a given direction, we can know with certainty where that particle will be at any time after it is thrown (as any rookie soccer player trying to score a goal learns). A wave, on the other hand, is like a disturbance that is present simultaneously in many locations in space. Unlike particles, waves can circumvent obstacles and superpose to yield a resultant wave that is the sum of the interfering waves. Such being the case, how is it possible that any entity is both a wave and a particle? As we already hinted at in the context of light, from the perspective of quantum theory this means that experiments may be conducted in which the particle nature of the system will be manifested, while other experiments may be conducted in which its wave-like nature will be revealed. If, for example, we aim and fire an electron gun at a partition with two slits and a screen behind them, an interference image of the electrons will be obtained on the screen that will look exactly like the interference image of a wave, regardless of which slit the electron gun is aimed at.

According to one interpretation, the wave-like nature of the system reflects the probability that the system will be in a certain place. In other words, the electron is not exactly a wave, but it has a probability of being in different locations in space at any given time, rather than only where we would expect to find it according to Newton's equations of motion. Each time we repeat the same experiment, we will discover the electron in a slightly different location; we cannot predict where it will be each time, but we can calculate the probability of finding it there.[3] Indeed, this probability has a wave-like nature and the distance to which the particle may deviate from the original path is approximately equal to the wavelength. According to this notion, if we aim a gun at the center of a target with absolute accuracy, the bullet discharged from the gun might well hit

---

[3] The philosophical idea underlying this interpretation is that there is no "quantum reality", but only a formula for calculating probabilities of measurement results. This is one of the interpretations Einstein opposed, claiming that "God does not play dice".

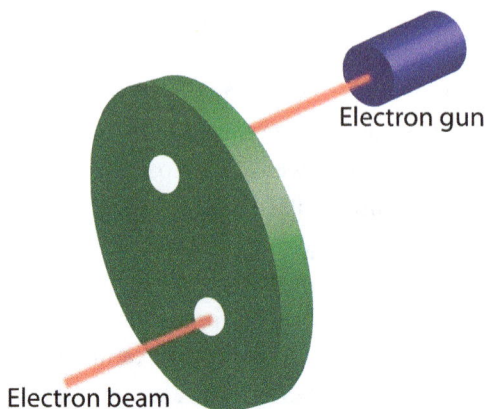

Figure 3.   An extreme demonstration of the uncertainty principle.

a completely different part of the target, despite the fact that that is not where the gun was aimed (see Figure 3 for illustration). This is what actually happens in the electron gun experiment described above. What then determines the wavelength, or alternatively, the extent to which any particle deviates from the expected trajectory? According to de Broglie, the answer is — its mass. The greater the particle's mass, the shorter its wavelength and the smaller its quantum deviation. For everyday objects with large masses, such as a tennis ball, a table, and even an ant, the wavelength is so short that its quantum deviations are indiscernible, even in the most accurate experiments. For tiny particles, however, like electrons and protons, the wavelength is long enough so that these deviations are measurable. Many different experiments conducted over the years established the wave–particle duality of physical systems, and showed in an almost certain manner that nature indeed suffers from schizophrenia.

Einstein's and de Broglie's works eventually led Werner Heisenberg to formulate one of the most basic principles of quantum theory, the *uncertainty principle*. This principle asserts that certain pairs of physical properties, such as momentum and location, cannot be measured simultaneously with absolute accuracy. For example, it is impossible to know with complete certainty both the momentum and the location of an electron. The more accurately we try to measure an electron's location, the greater will be the uncertainty regarding its momentum. And vice versa: if we attempt to measure its momentum with higher accuracy, the uncertainty regarding its location will increase. Similarly, it is impossible to accurately measure both the energy of a particle and the time at which the energy measurement is performed. For instance, if we install a detector that measures the energy of a colliding electron as well as the time at which the collision takes place, the more accurately we measure the collision energy, the greater will be the uncertainty regarding the time of the collision, and vice versa. The uncertainty principle applies to all kinds of particles and systems in nature, including light and radiation, and it does not reflect a difficulty in finding an appropriate measurement method but rather a basic property of quantum systems. In fact, it can be said that the system itself "does not know for certain" where it is, and what direction it is moving in, or as Arthur Eddington, one of the greatest astrophysicists of the early 20th century, said: "Something unknown is doing we don't know what." The formulation of Heisenberg's uncertainty principle, which on a philosophical level states that reality cannot be measured accurately, dragged the scientific community into a heated argument that continues until present day. Attempts have been made, and are still being made, to develop alternative quantum theories in which this principle will not be required, but at least on a practical level most scientists support the uncertainty principle.

5. The Exclusion Principle and the Periodic Table

Parallel to the development of the quantum theory, as described above, progress was also made in the understanding of the atomic structure. Experiments conducted by Ernest Rutherford (who started his professional life working under J. J. Thomson) and others began elucidating the structure of the atom, and the picture that emerged, as described previously, was of a tiny nucleus surrounded by electrons, like the solar system. It also was established that the distance between the electrons and the nucleus is much larger than the size of the nucleus itself. Thus, the atoms and the matter that is composed of atoms, including the air we breathe, the chairs we sit upon, the Earth and the Sun, are essentially a kind of web suspended in an empty space, which is permeated by electromagnetic forces that act between the electrons and the atomic nuclei. The main problem with this model, which was recognized relatively early on, was that atoms are unstable by nature. From the equations of the electromagnetic theory, it follows that electrons that revolve around nuclei must emit electromagnetic radiation and, as a result, they lose energy and collapse into the nucleus within less than a billionth of a second. This result contradicted experiments that showed that atoms are indeed stable over time. Another problem that emerged was that the model failed to provide a suitable explanation for the emission spectra of excited atoms, as were measured in multiple experiments. Such experiments revealed that each substance has a unique spectrum, that is composed only of certain wavelengths or colors (an example is a sodium lamp), which are called *emission lines*, because the spectrum looks like a collection of lines instead of being continuous. In other words, the spectrum of light emitted from atoms is discrete (quantum-like).

In 1913, Nills Bohr, a Danish physicist of Jewish origin, proposed a solution to these problems that combined the atomic model with the quantum theory as it was known at the time. Bohr's model (see Figure 4)

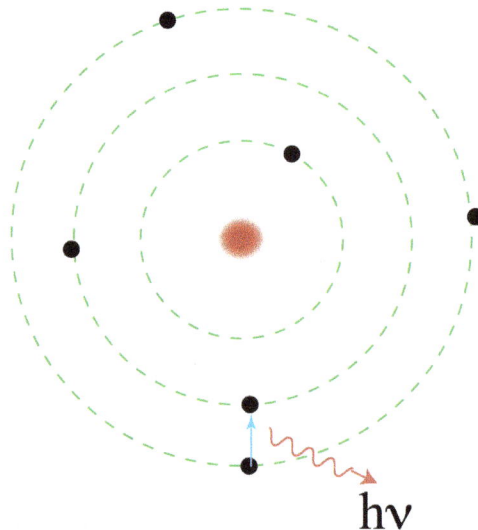

$$h\nu$$

Figure 4. Bohr's atomic model: Electrons (black circles) can exist only in certain orbits, indicated by the green dashed lines, and not between them. When an electron "jumps" from one orbit to the next, it emits a photon whose energy equals to the difference in energy between the two orbits.

assumed that electrons did revolve around atoms in a circular motion, but that they cannot be located anywhere (meaning, at *any* distance from the nucleus), but only in particular orbits whose angular momentum is an integer multiple of Planck's constant, *h*. In simpler terms, Bohr claimed that electrons can be located only in orbits with certain rotational velocities. Particularly, electrons are "forbidden" from coming too close to the nucleus. That is, there is one orbit that is the closest to the nucleus, and it is impossible for electrons to get any closer than that. De Broglie later showed that the orbits in which the electrons are permitted to be, according to Bohr's model, are those whose circumference is equal to an integer multiple of the electrons' de Broglie wavelengths. This may be likened to guitar strings that can only produce notes for which the length of the string equals an integer multiple of their wavelength. In this picture, an electron is like a standing wave in the atom. Bohr outdid himself when he proposed also that an electron can "jump" from one orbit to an adjacent orbit that is closer to the nucleus, releasing one quantum of light (photon) whose energy equals the difference in energy between the two orbits. The opposite process is possible as well; a photon of appropriate energy can collide with an atom and be absorbed by it, while "bouncing" an electron from an inner orbit to an adjacent orbit that is farther from the nucleus. Using the model he developed, Bohr calculated the spectra of various materials and obtained good agreement with the experimental data.

Bohr's model, which enabled electrons to jump from outer orbits to more inner ones, offered no explanation as to why electrons actually continue to move in different orbits and do not fall directly into the closest permitted orbit. Bohr assumed, without explanation, an atomic "shells" model whereby a certain shell (or orbit) cannot contain more than a finite number of electrons. When the shell is full, additional electrons can exist only in shells with larger radii, and electron transition is permitted only to shells that are not completely occupied. Thus, it was possible to explain, at least qualitatively, the existence of different elements (atoms) in nature as well as the periodic table of the elements: atoms of different elements differ in both the occupation number of their shells and the number of their populated shells. The regularity with which the elements are arranged in the periodic table and which gradually became clear in the early 20th century, drove researchers to seek a theory that would explain why a given shell can contain only a small number of electrons, as well as the exact number of electrons that can be arranged in each atomic shell. These attempts ultimately led Austrian physicist Wolfgang Pauli to formulate another important principle in quantum theory, namely *Pauli's exclusion principle*.

The exclusion principle states that only one electron can exist in any given quantum state (meaning, with any given energy and angular momentum). According to this principle, if people were electrons, identical twins could not possibly exist. Bohr's atomic model, which has since been refined, enabled scientists to calculate the number of different states possible in each shell. Since the exclusion principle permits only one electron in each such state, they were able to accurately determine the shell structure of each of the atoms that make up the elements in the periodic table. Thus, nearly 70 years after Dmitri Mendeleev introduced his periodic table to the world, a detailed explanation was offered for the classification of the chemical elements in nature.

The importance of Pauli's discovery may be summarized by the observation that the fact that two electrons cannot be in the same quantum state actually enables the existence of atoms, molecules, and life. If not for Pauli's exclusion principle, the electrons surrounding the atomic nucleus would crowd together in the center, the atoms would collapse into a very small volume, and chemical bonds, which are essential for the creation of molecules and more complex structures, would not be possible. The electric properties of materials, which determine whether a given material will be a good conductor, a semi-conductor, or an insulator, are also the outcome of Pauli's exclusion principle, as are the existence of mysterious stars called *white dwarfs*, with which we will become acquainted in the third part of this book. In general, it can be said that the existence of matter that has volume, as we are familiar with from our daily life, could not be possible without this basic principle of quantum theory.

The exclusion principle, which Pauli formulated in 1925, and its connection to the spin of the electron are, undoubtedly, Pauli's most well-known works for which he was ultimately awarded the 1945 Nobel Prize in Physics, following Einstein's recommendation to the prize committee. This, however, was not Pauli's sole contribution; his areas of interests as a talented physicist and scholar encompassed many subjects. His first article, which dealt with the theory of general relativity, was published when he was only 18 years old. At 21, he received his doctoral degree for his thesis addressing the quantum theory of molecular hydrogen. Another one of his revolutionary proposals pertained to the existence of a very strange particle called *neutrino*, which was mentioned earlier.

6. Antimatter

Every particle in nature has a counterpart with an identical mass and opposite electrical charge. This counterpart is called an *antiparticle*. Antiparticles may be thought of as mirror images of their corresponding "parent" particles. Just like every object placed on one side of the mirror has a counterpart that is its mirror image, so does every particle in nature have a kind of image in the mirror of nature that is represented by the antiparticle. The properties of a mirror image are identical to those of the original object, aside from its direction: the right-hand side of the object is reflected as the left-hand side of the image and vice versa. Similarly, for particles and antiparticles, the electric charge is the "reflection" property — the mass and other properties are identical in both, and only the sign of the electric charge is reversed. The antiparticle of the electron, alone, has been given a special name, the *positron*.

The possible existence of the positron was first raised by Paul Dirac, one of the most brilliant physicists of the 20th century and winner of the 1933 Nobel Prize in Physics, while seeking a relativistic quantum theory that would unify quantum mechanics and the theory of special relativity. Dirac's quest led him to one of the most earthshaking discoveries of the 20th century, namely the relationship between the electron's spin, the special theory of relativity, and the existence of antimatter.

A particle's spin represents its self-rotation, like that of the Earth around its axis. Every particle in nature has a kind of self-rotation, or spin; in the quantum theory, however, this rotation is more complex. When we set a top spinning on its axis, we can give it whatever rotational velocity we desire. In fact, we can determine not only its velocity, but also its direction — clockwise or counterclockwise. In quantum theory, rotation (or more precisely, angular momentum), like other quantities we've encountered, comes in discrete values. Not every rotational velocity is possible — only those for which the angular momentum is equal to whole- or half-integer multiples of Planck's constant, h, are permitted. A rotation that corresponds to an integer multiple of h is called an *integer spin*, while one that corresponds to a half-integer multiple of h is called a *half-integer spin*. To demonstrate this, let us assume that Earth is a particle that complies with the quantum theory, and let us assume that its current orbital period — once every 24 hours — represents one quanta. So, if Earth was a particle with an integer spin that complies with the quantum theory, then it could, in principle, revolve around its axis once every 24 hours or once every 48 hours or once every integer multiple of 24 hours, but not once every 30 hours, for instance. It could also revolve in multiples of 24 hours in the opposite direction (in which case the Sun would rise in the west and set in the east). And, if Earth was a particle with a half-integer spin, it would be able to complete one rotation in 12-hour multiples in either direction.

Particles with half-integer spins are called *fermions*, after the Italian physicist Enrico Fermi. Electrons, protons, and quarks, and in fact all particles that make up matter in the universe, are fermions. More accurately put, electrons and protons are spin-1/2 particles. This means that they are permitted only two values of self-rotation. If, in the above example, Earth was an electron, then it could revolve around its axis only once every 12 hours, in either clockwise or counterclockwise direction. Photons (light particles), on the other hand, are integer-spin particles, as are all other force-mediating particles (to be mentioned later on), which are named *bosons* after Satyendra Nath Bose who, together with Einstein, investigated their properties. What Paul Dirac showed was that the half-integer spin of electrons follows from the requirement that the quantum theory complies with the principles of the theory of special relativity. It can be said that the influence of special relativity is manifested, among other things, in the appearance of the half-integer spin among fermions and of the integer spin among bosons.

However, not only is the spin property implied by the unification of quantum mechanics and special relativity, but also the existence of antimatter. The discovery of antimatter initially concerned Dirac due to its strange properties; but all attempts to get rid of it failed. The new quantum equations derived by Dirac, which complied with Einstein's theory of special relativity, also implied the existence of both the electron's spin and antimatter. Several more years passed before the nature of antimatter was fully elucidated. In addition to its scientific importance, Dirac's work and the discovery of the electron's spin led to the development of semiconductor technology — a technology that is the basis for every modern electronic device: Walkman, video games, modern televisions, and computers, are all the result of Einstein's and Dirac's work.[4]

In 1932, about four years after Dirac's theoretical prediction, the positron was discovered in laboratory experiments conducted by Carl Anderson. Later on, experiments conducted in particle accelerators revealed the antiproton, the antineutron, and the antiparticles of most of the elementary particles. Matter composed of antiatoms, meaning a positron revolving around an antiproton, is called *antimatter*. Although the creation of antiparticles in accelerators has become a commonplace occurrence, only in 1995 did scientists at the Swiss CERN laboratories succeed in creating an antiatom for the first time. Theoretically, nothing prevents the world of matter we are familiar with from having a reflection in the form of the world of antimatter — antiplanets, antistars, and antigalaxies; however, to date, there has been no real evidence of it, even though small quantities of antimatter have been observed in various systems in the universe.

When antimatter comes in contact with ordinary matter, they destroy each other and release a huge amount of energy. This process, which is called annihilation, is approximately 1,000 times more efficient than energy production through nuclear fission, which is why antimatter engines are any space engineer's dream. Indeed, a single kilogram of antimatter can produce enough energy to power a car continuously for about 100,000 years. Various research groups, including the National Aeronautics and Space Administration (NASA), have been working for a while on the design of a space engine that is based on antimatter. One of the main challenges in the production of antimatter is its storage. It cannot be stored in ordinary containers or vessels, since any contact between the antimatter

---

[4] Stephen Hawking once said that Dirac would have made a fortune if he had patented his equation.

and the sides of the vessel will immediately result in uncontrolled annihilation and a tremendous explosion. One of the ideas being explored is an attempt to capture antimatter using magnetic fields so as to prevent contact with surrounding matter.

It should therefore come as no surprise that antimatter and the possibilities it holds have permeated also into science fiction books and movies. For example, the theft of antimatter from the Swiss CERN laboratories served as the basis for Dan Brown's novel, *Angels and Demons*, in which antimatter was to be used to destroy the Vatican during a papal conclave.

Antimatter is reserved an important place in the cosmological theories that describe the creation of the universe. In the early universe, close to the time of the Big Bang, matter and antimatter both existed in large quantities. Due to the special conditions that prevailed in the ancient universe, the existence of the two was possible without mutual annihilation. Then, the antimatter disappeared and all that remained was the matter.

7. The Four Forces

As far as we know, there are four elementary forces in nature: the gravitational force, the electromagnetic force, the weak force, and the strong force. These four forces of nature differ from one another in their essence and their properties, and remain separate in terms of the entities on which they act: gravity acts only between masses, the electromagnetic force acts only between electrical charges, and the strong force acts among quarks. Nevertheless, these forces seem like members of the same family, and in the past, used to be united.

Two of these forces are familiar to us all from our everyday life:

The gravitational force is responsible for falling of bodies toward the center of the Earth, for the revolving of planets around the Sun, for the capturing of stars by the galaxy, and for the expanding of the universe itself.

The electromagnetic force is responsible for most phenomena around us: attraction and repulsion between electrical charges, magnetism, transmission and reception of radio waves, sunlight, thunderstorms, and so on. The electromagnetic force also determines the properties of atoms and molecules, controls chemical processes, and dictates the structure of matter. In fact, some of the mechanical forces we encounter in our daily life result from the electromagnetic force too. Friction, for example, originates in the electrical force that acts between molecules at the contact points of the two bodies in question. The force that a spring applies when it is stretched also stems from an electrical force acting between the spring molecules.

The gravitational force and the electromagnetic force are long-range forces; in other words, they are discernable at large distances from their source. The electromagnetic force is, however, much stronger than the gravitational force. For example, the electrical force that acts within an atom, between the proton and an electron revolving around it, is 40 orders of magnitude (that means 10^{40} times) stronger than the gravitational force that acts between the two particles.

The two other forces are short range and are discernible only on sub-atomic scales:

The strong force is the force that prevails in nuclear physics. This force is the "adhesive" that holds the atomic nucleus together in the face of the protons' electrical repulsion, and without which atoms could not exist. These nuclear forces reflect the action of the strong force among the nucleon building blocks, the quarks.

The weak force relates to certain processes in the atomic nucleus, and in particular to the radioactive phenomenon familiar to us from nuclear reactors, atom bombs, and cores of stars. The weak force is also responsible for collisions between neutrino and other particles, as will be explained in Chapter 7.

Since these two forces act only over very short ranges, we do not often encounter them in our daily lives. To feel their effect, sub-atomic scales must be probed, an activity that requires a great deal of energy. This can be achieved in experiments conducted in large particle accelerators, where the effect of these forces is, indeed, observed.

8. Force Mediators

Why does an electron that is so far from a proton or from another electron feel an electrical force? Why do two distant masses exert a force on each other even though they are not even in contact? In other words, what mediates the forces between particles?

According to modern physics, particles are continuously being transferred between the electric charges in the case of the electrical force, between masses in the case of the gravitational force, and between quarks in the case of the strong force, and it is these particles that mediate the different forces. Each kind of force has its own special mediating particles.

In the case of the electromagnetic force, the mediating particle is the photon, the same light particle that appeared in Einstein's model of light, as mentioned previously. When two electrical charges, say an electron and a proton, are in the same vicinity, they sense each other by continuously emitting and absorbing photons (we can also say that they continuously emit and absorb electromagnetic waves).

In a similar manner, quarks communicate with each other by exchanging particles called *gluons* (from the word *glue*, meaning that the gluons glue the quarks to one another). A more detailed discussion of gluons is presented in Chapter 12. Gluons are a kind of electromagnetic radiation of the strong force. The weak force also has mediators, as does the gravitational force, whose mediators are called *gravitons*.

9. Spin and the Social Customs of the Elementary Particles

For completeness of the discussion, we will once again emphasize that, as opposed to the matter particles — electrons, quarks, protons, and neutrons — which are fermions, that is, have half-integer spins, the force-mediating particles — photons, gluons, and gravitons — are bosons, and have integer spins. It also turns out that these two types of particles have different social customs. Bosons are social particles; they like to crowd together. The significance of this tendency, in terms of the quantum theory, is that they do not comply with the Pauli exclusion principle, so that any system may be populated with an unlimited number of identical bosons. Low-temperature matter made of bosons has the tiniest volume. In contrast, fermions are reclusive, as we have seen already: only one fermion is permitted in any given quantum state, as defined by the Pauli exclusion principle. This is why the matter we are familiar with has considerable volume, even though it consists mainly of empty space.

10. Symmetry and Conservation Laws

The principle of symmetry plays a central role in physics in general, and in theories of elementary particles in particular. Symmetry is familiar to use in our daily life as a manifestation of beauty and aesthetics. According to the mathematical definition of symmetry, an object (or mathematical equation) is symmetrical under a certain transformation, if the transformation does not change it. The three basic symmetries in space are reflection, rotation, and translation. For example, the sequence OIO is symmetric under reflection, since when reflected in a mirror, it remains unchanged. The sequence OII, on the other hand, is not symmetrical under reflection, since its mirror image is IIO.

Symmetry plays an important role in biology and chemistry, and is especially important in the study of molecules. One interesting aspect is the difference between the chemical properties of two molecules that are mirror images of each other (a phenomenon called *chirality*), a difference that has significant implications in the chemistry of life processes.

In physics, symmetry has a central role. The conservation laws result from the symmetry of the laws of nature, or more precisely, of the basic equations that describe these laws. For example, the law of conservation of linear momentum is related to translational symmetry while the law of conservation of angular momentum is related to rotational symmetry. The law of conservation of energy is associated with translational symmetry of time. A simple example of this relationship is a particle moving at a constant velocity. If we measure the particle's velocity at any given time, we will obtain the same result. Hence, the velocity of the particle does not change when time "moves" from "early" to "late", or vice versa. And in more formal terminology, the particle has symmetry under time translation. On the other hand, the fact that the particle's velocity is constant means that no forces are acting upon it, hence its kinetic energy is conserved. In comparison, if the particle is accelerating, there is no time translation symmetry since the velocity we will measure at a certain time will be smaller than the velocity that will be measured at a later time. In this case, the particle's kinetic energy increases over time and is, therefore, not conserved.

In the early 20th century, a German mathematician named Emmy Noether formulated and proved the following theorem: "Every symmetry in nature has a corresponding conservation law, and every conservation law is associated with a symmetry in nature". This theorem, which is called *Noether's theorem*, has become one of the most fundamental principles of modern physics, and has many implications. For instance, the law of conservation of electric charge results from internal symmetry (which is a kind of generalized rotation symmetry) of the electromagnetic field called gauge invariance. Another example is the classification of hadrons — subatomic particles that are made of quarks (Chapter 12, Figure 1) — which is based on a symmetry referred to as the *Eightfold Way*,[5] discovered by Murry Gell-Mann and Yuval Ne'eman in the 1960s. The fact that a proton will never ever become an electron also stems from the internal symmetry of the elementary particle system. Several more conservation laws exist, which were inferred from thousands of experiments in particle accelerators conducted over the years, and that result from internal symmetries of the other forces of nature.

Experience tells us that despite the important role of symmetry in nature, many phenomena reflect asymmetry. It suffices to look around us in order to sense it. Asymmetry is felt even at the microscopic level. Thus, for instance, the masses of the elementary particles are not identical: protons are much heavier than electrons, and even the quarks themselves have different masses. There are other examples as well. One of the greatest enigmas is the asymmetry that exists in the universe between matter and antimatter. If all matter in the universe was created from energy, as the Big Bang theory claims, then we would expect the universe to contain equal amounts of matter and antimatter. But in fact, we are witnesses to the existence of only matter (aside for minute quantities of antimatter, created in violent processes in the universe, and as the result of collisions of cosmic radiation and matter in the galaxy).

An even more extreme example is time itself. All processes in nature occur in only one direction in time — from the past to the future. There is no absolute symmetry in time. Nobody is born old and becomes younger over the years (like a movie played in reverse), clocks all advance in the

---

[5] Gell-Mann, who is known for his fondness for unique names, borrowed the term *Eightfold Way* from Buddhism.

same direction, we remember the past (not the future), and even the universe evolves in only one direction in time.

What, then, is the reason for this lack of symmetry in a world that allegedly was supposed to be symmetrical? As we will soon see, it stems from *symmetry breaking* — a phenomenon in which a symmetrical system suddenly loses its symmetry. The role of symmetry breaking in modern physics is no less important than that of symmetry itself. As we will see, according to modern perception, the universe was highly symmetrical in the initial stages following its creation, but this symmetry increasingly broke as the universe increasingly expanded and became cooler.

11. Symmetry Breaking

To understand what the term *symmetry breaking* actually means, consider the system depicted in Figure 5. First, let us assume that the red sphere is placed exactly on the tip of the protrusion, so that it is balanced and does not fall off (left). This system clearly has symmetry. For instance, if we rotate the entire system around the axis indicated by the dashed line, no change will be discernible. It is clear that in this state, the system also has reflection symmetry. It is also evident that the red sphere is in a highly unstable state. Any disturbance, slight as it may be, will cause it to drop into the depressed ring, as illustrated on the right. In this state, symmetry is lost. If we rotate the system in the right-hand image by 180°, the sphere will now be at the left of the protrusion. It can similarly be seen that the system has lost its reflection symmetry as well. This example demonstrates that the slightest deviation from a perfectly symmetrical state can cause the

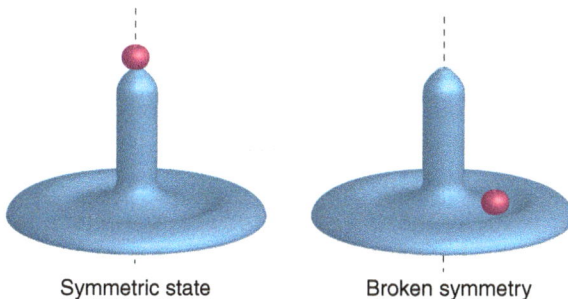

Symmetric state   Broken symmetry

Figure 5. Symmetry breaking.

symmetry to break. In addition, when the sphere was on the tip of the pro-trusion, it had an equal chance of dropping in any direction, but it sponta-neously (meaning, in an unplanned manner) "chose" to drop in one specific direction. Such a process is called *spontaneous symmetry breaking.*

Spontaneous symmetry breaking occurs in many natural systems. One such example is a magnet: at high enough temperatures, ferromagnetic (iron-like) matter is not magnetized. If we heat an ordinary magnet to a high enough temperature, it will lose its magnetization. In this state, the atoms within the material itself have no preferred direction, and there is symmetry. Indeed, if we rotate the magnet or reflect it, the picture of the atoms within it will not change. If we cool the magnet to a low enough temperature, a magnetic field will suddenly appear. In this state, the atoms are oriented in a preferred direction, the magnetic pole, and some of the symmetry is lost. In fact, the atoms, which were randomly oriented when the temperature was high, all suddenly align themselves in the same direc-tion after the magnet is cooled. The symmetry in this example is spontane-ously broken during the magnetization process.

The above examples are not unique. Symmetry breaking occurred and is still occurring in many natural systems, on both the macroscopic and microscopic levels. And as we shall see now, the most important and inter-esting example has to do with the forces of nature.

12. Unification of Forces and the God Particle

According to the "standard model", at high enough energies, the elec-tromagnetic force and the weak force, which supposedly account for completely different phenomena, are unified into one single force. In this state, the system's symmetry is higher and, according to the unifica-tion theory, it can be shown that this symmetry requires all particles, including electrons, positrons, quarks, and force-mediating particles, to be massless. Where then does the mass of these particles suddenly come from? The appearance of mass is related to the breaking of symmetry and to another particle, the *Higgs boson.* This particle was named after physicist Peter Higgs, who together with Robert Brout and François Englert, proposed its existence in the 1960s. This particle has no special role at high enough energies, when symmetry is maintained. When the energy drops below a certain threshold, however, sudden symmetry breaking occurs and quasi-condensations of the Higgs particle appear, like the formation of clouds in the sky when the humidity rises and the temperature drops, or like the appearance of water droplets on a car windshield when the temperature of the outside air drops suddenly. These condensations affect the other particles and lead to the appearance of mass. Imagine that electrons, quarks, and the other particles, which were massless and free before, are suddenly placed in a viscous fluid (those Higgs particle condensations). When they move about in this fluid, they feel its resistance to their movement — this resistance is equivalent to mass.

The unification theory described above is called the *standard model*. Thousands of experiments conducted over the past decades have confirmed many predictions of this model, although some open questions still remain. The detection of the Higgs particle presented one of the most tremendous challenges in experimental particle physics. The two main reasons for this are the very high energy required for its detection, which requires an especially large particle accelerator, and its extremely short lifetime. The large hadron collider (LHC) located at CERN Laboratories in Switzerland was built especially for this purpose. On 4 July 2012, CERN announced the detection of a new particle, consistent with the Higgs boson predicted by the standard model. On 8 October 2013, the Nobel Prize in Physics was awarded jointly to François Englert and Peter W. Higgs for their theoretical work on the Higgs mechanism. In subsequent experiments since its discovery, this new particle has been consistently shown to behave in many of the ways predicted by the standard model.

Tremendous efforts are being invested in attempts to develop a complete unified theory in which the notion of unification will apply to all forces of nature — strong, weak, electromagnetic, and gravitational. These attempts have so far led to the development of the *supersymmetry theory* and the *superstring theory*, but to date, neither of these theories has gained any empirical confirmation and they are still considered pure speculation. In fact, it is not even clear which of the many models that enable the unification of the forces is correct. Indeed, not everyone even agrees that the unification of all forces is necessary, although existing evidence indicates that at least two of the forces — the weak force and the electromagnetic force — were once unified.

When exactly did the symmetry of the forces in nature break? According to the standard model, the early universe was characterized by nearly perfect symmetry. All particles were massless and all of the fundamental forces were unified to one single force. The universe also contained equal amounts of matter and antimatter. A very short time after the Big Bang, the universe cooled down and symmetry broke, similar to symmetry breaking in ferromagnetic matter, as described above. Following this phase, the electrons, quarks, and remaining particles acquired their mass; the only force that prevailed until then split up into the four forces we are familiar with today, and most of the antimatter in the universe was annihilated. Thanks to this symmetry breaking, matter and life were created, and thanks to it human beings exist as well.

Chapter 5

Principles of Modern Astronomy

"Astronomy's much more fun when you're not an astronomer"

— Brain May

Since the invention of the telescope in the early 17th century and Galileo Galilei's use of it for astronomical observations, the science of astronomy has undergone a true revolution. Not only a technological revolution that has led to the construction of giant telescopes and measuring devices of unprecedented accuracy, but rather a revolution in the perception of astronomy. Ordinary telescopes that detect visible light constitute only a small fraction of the means available to modern astronomers for studying the universe and its secrets. Radio waves, X-rays, gamma-rays, cosmic radiation, neutrinos, gravitational waves, and antimatter are just several of the wide variety of radiation and particles used to investigate the universe in modern times. The revelation of this plethora of information carriers (sometimes referred to as *multi-messenger information*) requires different and diverse methods and sophisticated instruments installed underground, on land, and in space. These various detection methods are based on principles of modern physics and require an understanding of the most basic physical theories. In the following chapters, we will review the different kinds of radiation used to explore the universe and its secrets, and the ways in which they were discovered. We will also mention some of the experiments that are at the forefront of astronomy today, and are being conducted at various locations around the world to study the phenomena which will be discussed in the Third Episode of this book.

Chapter 6

Types of Electromagnetic Radiation

Electromagnetic radiation is emitted from accelerating electric charges and is transmitted through various media (including absolute vacuum) by means of changes in the electric and magnetic fields in space. This is the most well-known kind of radiation, and our life without it is hardly imaginable. Radio and television broadcasts, cellular phones, sunlight, electric lights, X-ray imaging, infrared devices, microwave ovens, and radar are all just a few examples of electromagnetic radiation and its applications. And as we mentioned earlier, the discovery of electromagnetic radiation also led to the development of Einstein's theory of special relativity. The velocity of an electromagnetic wave in vacuum is the highest in the universe and one of the basic constants of nature. Its value is 300,000 km/sec (or, to be precise, 299,792 km/sec), and its common symbol in scientific literature is c. This velocity, which is called *the speed of light*, is identical for all kinds of electromagnetic radiation — radio waves, microwaves, visible light, ultraviolet radiation, X-rays, and gamma rays. The parameter that distinguishes between these types of radiation is the wavelength, or alternatively, its frequency.[1] Other than this difference, all kinds of radiation mentioned are identical. Visible light, the range of wavelengths to which the human eye is sensitive, constitutes only a very small portion of the electromagnetic spectrum. Radio waves have the longest wavelength, reaching several kilometers, while gamma rays have the shortest wavelength — less than one millionth of a nanometer (one nanometer equals

---

[1] The product of a wavelength and its frequency equals the speed of light.

Wavelength (meters)

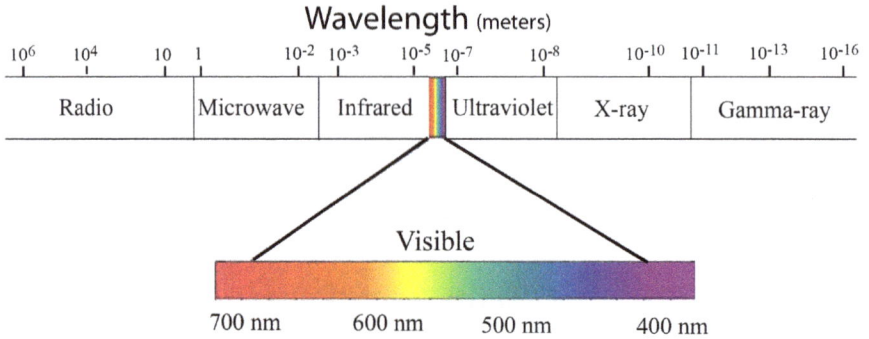

Figure 1. The electromagnetic spectrum: The different kinds of radiation are distinguished from one another by their wavelength. Visible light consists of the range of wavelengths to which the human eye is sensitive. The color of the visible light is determined by the wavelength: violet is located in the range of the shortest wavelengths of visible light, whereas red is in the range of longer wavelengths. The symbol nm stands for nanometer, a billionth of a meter.

one billionth of a meter). The possible existence of electromagnetic waves was first predicted in the late 19th century by British physicist James Clerk Maxwell, after formulating the laws of electromagnetism as four basic equations called *Maxwell's equations*. These equations describe every phenomenon related to electric and magnetic fields, including phenomena related to electromagnetic waves. In 1887, Heinrich Rudolf Hertz built the first radio wave transmitter and receiver, demonstrating the existence of electromagnetic waves. Indeed, Hertz's experiments led to the development of the telegraph and radio devices, and the frequency unit, which is measured in cycles per second, is named after him. A frequency of one megahertz (one million hertz) is equivalent to a wavelength of 300 m.

As mentioned, electromagnetic radiation is a wave-like phenomenon. A wave may be defined as a disturbance that propagates through space and is simultaneously present in multiple locations. Most kinds of waves we are familiar with propagate in some sort of medium. For instance, sound waves are variations in pressure and density that propagate in air (or in water or in any other medium in which they are created). If we were to turn on a loudspeaker in outer space, where there is almost absolute vacuum, we wouldn't hear a thing, since sound waves cannot propagate through vacuum. When we cast a stone into still water, a change is created in the height

What is a Spectrum?

A spectrum is the composition of frequencies contained by a given wave. In the case of visible light, the spectrum reflects the wave's color composition. Sunlight has a broad spectrum, which contains a continuous sequence of colors. A sodium lamp, on the other hand, has a narrow spectrum that contains only two adjacent frequencies that correspond to two similar hues of orange. The bright color we see when we look directly at the Sun is, in fact, the sum of all of the colors a human eye can see. Light can be separated into its different colors by refraction, for example, using a prism (see below) or a drop of water, since waves of different frequencies have different refraction angles, depending on the wave's frequency. The most common demonstration of this principle is a rainbow, which is created when sunlight undergoes double refraction in water droplets.

Similarly, the spectrum of a sound wave is the sum of the sound frequencies the wave carries. Low frequencies correspond to bass tones, while high frequencies correspond to soprano tones. Playing the sixth note of the fourth octave of a piano will create a sound wave with a frequency of 440 Hz, which to a human ear will sound like an A note. Playing several notes simultaneously will create a sound wave with a broader spectrum, which is composed of the collection of different frequencies produced by each of the notes. The human ear will hear this composite wave as a chord. Random noise has an even broader spectrum, containing a continuous sequence of all frequencies.

of the water surface at the point of the submersion of the stone, which propagates immediately, in all directions, along the surface of the water.[2]

Similarly, electromagnetic waves are changes in the electric and magnetic fields that propagate from the wave source throughout the entire space. The source of such a wave can be, for instance, a radio station transmission antenna, the microwave generator in our kitchen microwave oven, or atoms in the sun that emit the light we see. As opposed to most of the waves we are familiar with, which need a medium in which to propagate, electromagnetic waves can propagate even in absolute vacuum thanks to reciprocal changes in the electric and magnetic fields. This fact was not obvious when electromagnetic waves were discovered in the 19th century. At that time, scientists, who were already familiar with sound waves, sea waves, and waves created when a string is plucked believed that light propagates in a kind of medium referred to as *ether*. The nature of ether was not clear, and unsuccessful attempts were made to measure it. A breakthrough came following a famous experiment conducted by two researchers, Albert Abraham Michelson and Edward Morley, who published their findings in 1887. Michelson and Morley's experiment proved beyond any doubt that ether does not exist, and that the speed of light does not depend on the motion of the source relative to the observer (the modern formulation, given by Einstein, is that the speed of light is the same in all reference frames). This experiment was one of the motivations for the development of Einstein's theory of special relativity.

Despite differences in the kinds of waves and media in which they propagate, all waves are described, mathematically, by the same equation — the *wave equation* — and are therefore subject to the same basic rules.

1. Interference and Diffraction of Waves

Two phenomena that are unique to waves are interference and diffraction. When waves are transmitted from two different sources, for instance, two

---

[2]These waves are not sound waves, and are sometimes called "surface waves" or "ripples". While sound waves represent changes in the pressure and density of the medium itself, surface waves represent changes that propagate along the interface between different media (air and water in this example). In fact, when a stone is cast into water, both surface waves and sound waves are formed, with the latter propagating into the depth of the water.

The Speed of Light in Vacuum as a Universal Constant

To clarify the meaning of the statement "The speed of light is the same in all reference frames", imagine the following thought experiment: You are in a spacecraft traveling at the constant velocity of 100,000 km/sec relative to your friend on Earth. The spacecraft is equipped with a laser gun and a device that can measure the velocity of the laser beam. Your friend on Earth is holding an identical measuring instrument. At a certain point in time, you shoot a laser beam in the direction in which the spacecraft is traveling. Your measurement shows that the laser beam is traveling at a velocity of 300,000 km/sec relative to the spacecraft. What will be the beam's velocity according to your friend's measurement?

Since you shot the beam in the spacecraft's direction of motion, you are certain that your friend's result will be the sum of the two velocities, in other words, 400,000 km/sec. When you return to Earth, however, you are surprised to discover that your friend's measurement result was identical to the velocity that you measured on the spacecraft, 300,000 km/sec. You conclude that there must have been a mistake in the measurements and you repeat the experiment. This time you direct the laser beam in the direction opposite to the direction of motion of the spacecraft. But to no avail; you obtain the same results again: the velocity your friend measures is exactly the same as the velocity that you measure. After many trials, you conclude that the speed of light is identical for all observers, and is always 300,000 km/sec, even if the light source and the observers themselves are moving with respect to one another.

If light traveled in a medium, like scientists at the end of the 19th century believed it did, then the velocity measured by an observer moving away from the light source should be different than the velocity measured by an observer moving in the opposite direction, that is, toward the light source. This is the difference that Michelson and Morley tried to measure in their famous experiment. The results showed that, in fact, there is no difference, like in the example of the two friends described above. After digesting their results, the physicists concluded that the speed of light is a universal constant! That means that the speed of light is identical for all observers, regardless of their own velocity relative to the light source. We again emphasize that this applies only to the speed of light in vacuum. Now, imagine that you are in a spacecraft, chasing a ray of light that is

(Continued)

(*Continued*)

traveling in space. You accelerate the spacecraft in order to catch up with the light, but since the speed of light is independent of the movement of the spacecraft, it is not possible — the light continues to "outrun" you, traveling at the same speed, regardless of the velocity you accelerate the spacecraft to. Hence, it follows that the speed of light is the highest velocity in the universe, since otherwise it would be possible, at least theoretically, to catch up with the light ray.

The explanation, according to the theory of relativity, of these strange results is that the length measured by rulers and tape measures and the time that passes on clocks and watches are those that depend on the movement of the observers, and they are constantly changing so that the speed of light, in other words, the ratio between the distance that the light travels and the time it takes to travel that distance, is the same in all systems.

radio antennas located at a certain distance from each other, a resultant wave is formed that is composed of the sum of the two waves emitted by the sources. This phenomenon is called *interference*. In certain places in space the amplitude of the resultant wave will increase (*constructive interference*), and in other places it will decrease (*destructive interference*). Waves, as opposed to particles, can also circumvent obstacles; this phenomenon is called *diffraction*. Diffraction is the reason we can hear a sound source even when we are hiding behind a wall — the sound wave simply circumvents the wall. In fact, interference occurs also in every diffraction process. For instance, when a wave circumvents several obstacles, a new wave front is formed behind each of the circumvented obstacles, which propagates in space. The waves formed behind the different obstacles interfere and form a resultant wave. Diffraction through a single slit (or aperture) in an opaque partition creates new wave fronts at each point along the slit, which interfere in space, as demonstrated in Figure 2. Figure 3 shows the diffraction pattern created by a circular aperture.

Diffraction can be used to separate light into its component colors since the interference pattern of the diffracting waves depends on wavelength. An instrument that does just this is called a diffraction grating and is used to measure the spectrum of light sources. Diffraction gratings are a

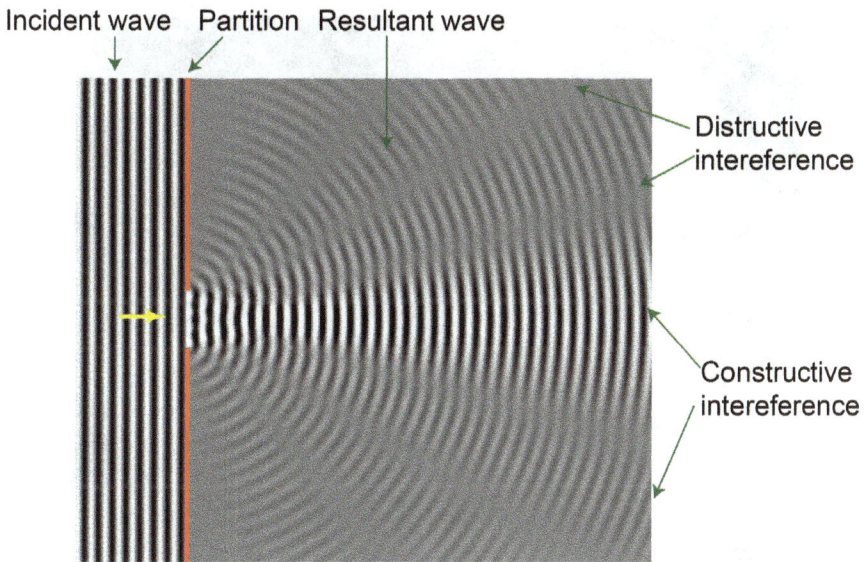

Incident wave   Partition   Resultant wave

Distructive
intereference

Constructive
intereference

Figure 2. Diffraction of a surface wave in a wave bath. A new wave front is formed at each point along the slit in the red partition. Interference of these waves creates a resultant wave pattern to the right of the partition. Where the color is solid grey, the interference is destructive and the resultant wave cancels out. Where there are stripes, the resultant wave is at maximum intensity. Courtesy of Richard F. Lyon, via Wikipedia.

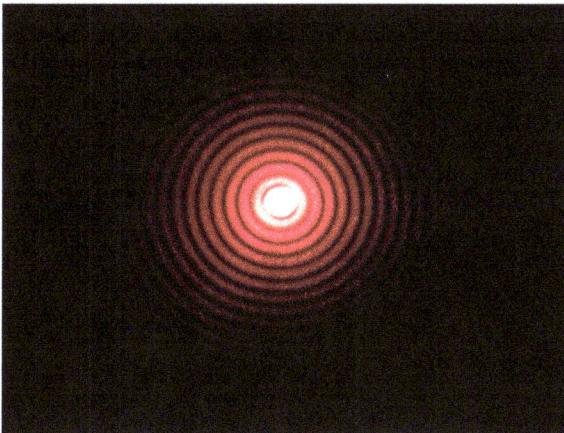

Figure 3. The diffraction pattern created by a laser beam traveling through a tiny circular aperture in an opaque partition. The aperture is at the center of the white spot.

Figure 4. The fine grooves of a CD can serve as a diffraction grating. The photo shows sunlight being reflected from the grooved surface of a CD. The location of the maxima and minima of the resultant wave amplitude are different for each color, or to be precise, for each wavelength, and so we can see the separate colors.

component of every modern telescope. A common CD, like those found in any household, can serve as an impromptu diffraction grating (Figure 4).

Many phenomena that occur around us are connected to the interference and diffraction of different kinds of waves, although we are not usually aware of it.

2. Reflection, Refraction, and Dispersion

Reflection and refraction of waves occur upon transitioning from one medium into another medium with different properties, as for instance, from air to glass. Such phenomena are familiar to us from our everyday life. Common examples are the reflection of light rays from a mirror and the distortion of images in water, as seen in Figure 5, a distortion caused by the refraction of light rays when transitioning from water to air.

Figure 5. Light rays reflected off parts of the pencil that are emerged in water are refracted when they transition from water to air, creating the bcroken-pencil effect seen in the photo.

In fact, reflection and refraction are two aspects of the same phenomenon. When a wave passes from one medium to another, like from air to glass, part of it is reflected and the other part enters into the other medium and is refracted (see Figure 6). The answer to the question what part of the wave is reflected and what part is refracted depends on the properties of the medium and on the wave's angle of incidence. An ideal mirror is a medium that reflects the wave in full. In contrast, a "transparent" material is a medium that transmits the entire wave and does not reflect back any of it. Most materials in nature are somewhere between these two extreme cases. Some materials absorb part of the refracting wave while traveling through the medium into which it entered; the energy of the absorbed wave turns into heat or some other form of energy.

Reflection and refraction result from the fact that in non-vacuum media (for instance, glass, water, or air) the velocity of an electromagnetic wave is lower than its velocity in vacuum. The reason for this is that the wave's electrical field excites the electrons in the medium, and in response, they create an electrical current that "delays" the wave's progress. Another

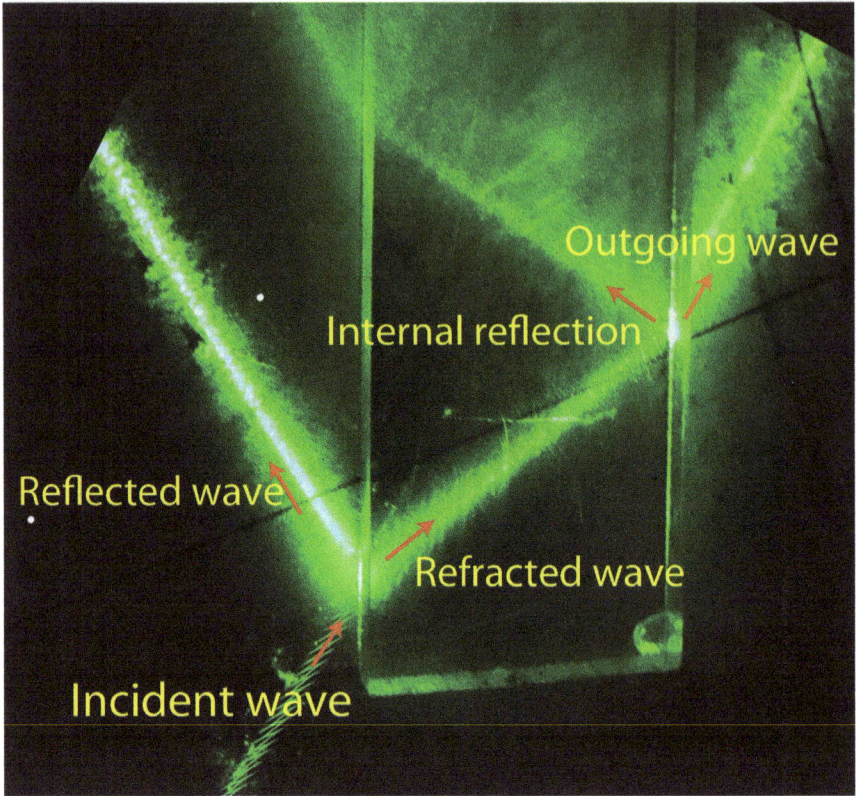

Figure 6.   A photo of the transition of a laser beam from air to glass. Part of the incident wave is reflected and part of it penetrates the glass and is refracted in it. The reflection angle is always equal to the wave's angle of incidence. The refraction angle, on the other hand, depends on the incidence angle and on the ratio between the refractive indexes of air and glass. Upon passing from glass back into air (at the right-hand side of the medium), a similar process occurs — part of the wave is reflected back into the medium and part of it is refracted outward. Photo courtesy of the Physics Dept. Demonstrations Lab, Tel-Aviv University.

way of saying this is that within a non-vacuum medium, electromagnetic waves do not travel freely, but are absorbed and emitted over and over again throughout their movement, and the rate at which this process of absorption and emission cycles progresses through the medium is slower than the speed of light in vacuum. The ratio between the velocity of the wave in vacuum and its velocity in a given medium is called the *refractive index*, and is denoted *n*. For instance, under standard conditions, the refractive index of air is 1.00029, of water — 1.33, and of glass — between 1.52 and 1.7, depending on the kind of glass. The speed of light in air is, therefore, very close to its speed in vacuum (about 99.97%), while in water it is only about 75% of its speed in vacuum, and in glass, even lower. The refractive index is the parameter that characterizes the optical properties of a medium; reflection and refraction occur upon transition of a wave from a medium with a certain refractive index to a different medium with a different refractive index. The ratio between the wave's angle of incidence and its angle of refraction depends only on the ratio between the refractive indices of the two materials. It can be shown that the angle of refraction is such that the time that it takes the wave to travel between two points — one within the medium from which the wave exited and the other within the medium into which it entered — is the shortest.

In most materials, the refraction angle depends on the frequency of the incident wave. In the case of light, for instance, the refraction angle of a red wave is smaller than that of a blue wave. As a result of this difference, waves of different frequencies separate (or disperse) when the wave is refracted. This phenomenon is called *dispersion*. Figure 7 presents an example of dispersion in which a beam of white light penetrates a glass prism and separates into the different colors it is composed of. If a screen is positioned behind the glass prism in which double refraction of the light is occurring, like in the photo, the colors will get separated on the screen and we can measure the color composition, or in professional terms, the spectrum of the original light ray.

3. Cherenkov Radiation — An Optical Sonic Boom

The fact that the speed of light in a medium is lower than the speed of light in vacuum enables the emission of type of radiation called *Cherenkov radiation*. This radiation is actually electromagnetic radiation that is emitted when a charged particle travels at a velocity that exceeds the speed of light in that same medium (hence, Cherenkov radiation cannot be emitted in vacuum). The movement of the charged particle excites the molecules of the medium, and these emit the electromagnetic wave in response. Since the velocity of the particle in the medium is greater than that of the wave, an effect is created that is similar to a sonic boom created by an aircraft traveling through the air at a speed that exceeds the speed of

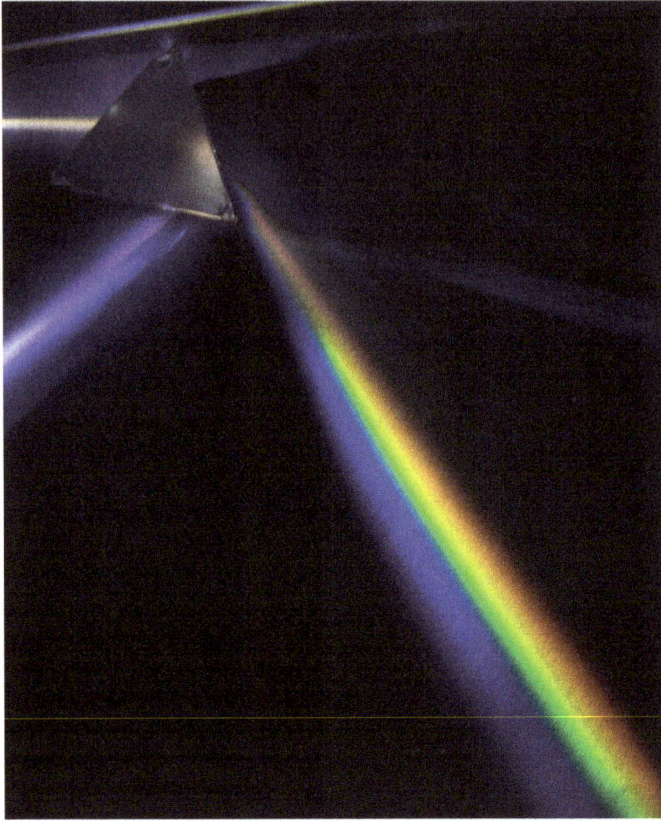

Figure 7. Dispersion in a prism: The beam of white light (top left) penetrates the glass prism. Upon leaving the prism (right), the light beam separates into the colors that compose it. The second beam seen in the photo (bottom left) results from the internal reflection upon transition from the prism to the air. Courtesy of D-Kuru/Wikimedia Commons.

sound. Indeed, it can be said that Cherenkov radiation is a kind of electromagnetic shock wave (see Figure 8).

This effect is named after Russian physicist Pavel Alekseyevich Cherenkov who discovered the phenomenon in 1934. Cherenkov noticed a bluish light glowing from a bottle of water that was bombarded with high-speed particles generated by radioactive decay. Cherenkov earned the 1958 Nobel Prize for Physics for his work, and his gravestone at the Novodevichy Cemetery in Moscow bears the illustration of the process he discovered (see Figure 9).

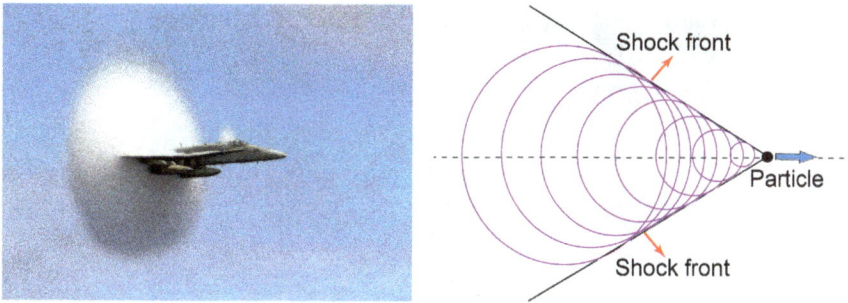

Figure 8. The diagram on the right illustrates the Cherenkov radiation process. The charged particle "outruns" the emitted waves, and a "shock" front of light is created, similar to the shock wave formed when an aircraft exceeds the speed of sound. For the sake of comparison, the photo on the left depicts a shock wave created by an F-18 US Air Force jet. The condensation cloud enveloping the aircraft is formed as a result of the shock wave.

Figure 9. Cherenkov's gravestone.

Figure 10.   The blue light in the photo is Cherenkov radiation that is emitted from the water of a pool in which fuel rods of a nuclear research reactor are immersed. Had the reactor core been immersed in vacuum, no light would be emitted. Courtesy of Argonne National Laboratory.

In both air and water, Cherenkov radiation is emitted in the visible light range, and most of the intensity is concentrated in wavelengths that correspond to blue and violet. The blue color glowing from nuclear reactor pools originates from Cherenkov radiation (see Figure 10), where the light stems from the movement of high-speed electrons emitted during the fuel fission process that takes place in the core of the reactor. The light intensity is indicative of the rate of the reaction in the reactor.

Measuring Cherenkov radiation is one of the more common ways of detecting charged particles in particle accelerators. This method is also common in high-energy astronomy. As we shall see later on, certain gamma ray detectors, neutrino telescopes, and cosmic ray detectors are all based on the Cherenkov effect, and this constitutes the main reason for mentioning the phenomenon in this book.

4. Waves, Particles, and Detection Methods

All of the effects discussed above (reflection, refraction, and dispersion) are related to the wave-like property of electromagnetic radiation. In Chapter 4, Section 4, we explained that all physical systems, and particularly electromagnetic radiation, have a dual character — they behave as both wave and particle. Light in particular may be regarded as being composed of a collection of massless particles called photons. The wave-like nature of light is revealed in certain experiments, like those described above, while other experiments uncover its particle-like behavior. The duality property is especially manifested in the designing of detectors and measuring instruments.

Although all kinds of radiation mentioned here are part of the electromagnetic spectrum, each range of wavelengths requires different and unique detection methods. This stems from the fact that the interaction of different materials with electromagnetic radiation depends on the radiation's wavelength. For instance, the mirror of a regular telescope reflects visible light with high efficiency, but does not reflect X-rays and gamma rays. The reason is that the wavelength of these kinds of radiation is shorter than the average distance between the mirror atoms, which enables high penetrability. Similarly, a satellite dish is a good mirror for radio waves, but not necessarily for visible light. Furthermore, some of the detection methods take advantage of the wave-like properties of the radiation, like in visible light telescopes, radio antennas, and diffraction gratings used to measure spectra, while other methods exploit the particle nature of radiation, like the photoelectric effect used in photo-multipliers[3] and Geiger counters and ionization chambers used to detect X-rays and gamma radiation.

---

[3] A photo-multiplier is a device that enables the measurement of very weak signals of visible light and ultraviolet radiation. The device is based on the repeated emission of electrons by which the photoelectric current caused by the incidence of light is increased (or multiplied) a billion-fold.

In addition to the difficulty arising from the fact that each wavelength range requires different and unique detection methods, astronomy adds yet another challenge. Some types of radiation — ultraviolet, X-rays, gamma rays, and part of the infrared spectrum — are absorbed by the atmosphere (protecting us from the damages of such radiation), and in order to measure such radiation, we must launch balloons and satellites carrying appropriate detectors.

The two main requirements a telescope or any other astronomical instrument must fulfill are sensitivity and resolution, both of which must be as high as possible. The higher the sensitivity of the device, the paler and more distant the objects it is capable of discerning; higher resolution yields sharper images, which enables in distinguishing more details regarding the source. The sensitivity threshold of an astronomical instrument is proportional to the total surface area of all of its apertures. In a regular telescope, this means the surface area of the mirror; in a radio antenna array, this is the total surface area of the radio dishes; and in a gamma ray detector, this is the surface area of the gas tank that absorbs the gamma radiation. The angular resolution of a telescope or other detector is defined as the smallest angle at which two point sources can still be clearly discerned. At smaller angles, the two sources will appear as one blurry spot. The best angular resolution theoretically obtainable in any instrument is, to a good approximation, equal to the ratio between the wavelength of the measured wave and the baseline of the instrument (the baseline is the maximum distance between two light-receiving parts of the device). Sometimes, however, additional limitations exist that prevent realizing the instrument's maximum potential. For example, the resolution of ground-based telescopes is restricted by atmospheric turbulence that causes twinkling. This is why most large telescopes throughout the world are located in high geographic areas where the atmosphere is relatively thin. The atmospheric distortion can be completely avoided by launching the telescope into space or, alternatively, by using adaptive optics methods, as will be explained below.

We will now review the different aspects of radiation, beginning with the longest wavelengths and working our way through the spectrum all the way to gamma rays, which have the shortest wavelengths. We will also discuss some of the currently available means used to detect and measure each kind of radiation.

5. Radio and Microwave Astronomy

Radio emission covers the broadest range of wavelengths in the electromagnetic spectrum, ranging from hundreds of meters in the AM transmission range to several meters in the FM range. The range of wavelengths between several millimeters and dozens of centimeters, which is reserved as a window for cellular phone transmissions, microwave ovens and so on, is called microwave radiation.[4] Many objects in the universe, like the Sun and other kinds of stars, galaxies, pulsars, quasars, microquasars, and supernovae, which we will address later on, emit radio waves that carry unique and important information. While common, everyday applications translate the information carried by radio waves and received by radio receivers into audio signals, in astrophysical applications, radio waves are converted into images. Although radio waves have the longest wavelengths of all radiation types, techniques have been developed over the years that support the construction of giant devices with a better angular resolution than can be achieved with other kinds of radiation, including visible light. Thanks to such devices, radio images offer scientists a great many details at high levels of clarity.

Immediately after Hertz discovered electromagnetic radiation, hypotheses were put forward according to which the Sun and other astrophysical sources in the universe emit radio waves. Initial experiments with radio astronomy were conducted as early as the 1930s, but the big breakthrough in this area came after World War II, following the rapid advances in radio and radar technology. The first astronomical radio source was discovered by chance in 1930 by an engineer named Karl Jansky, who worked for Bell Telephone Laboratories. Jansky was investigating random sources of static noise that was interfering with radio transmissions, when he noticed that one of the sources was exhibiting cyclic modulations. Since the intensity of the source changed every 24 hours — the rotation period of Earth — Jansky suspected that the source was the Sun (like day/night changes). After a short investigation, this possibility was rejected and the source was found to be a distant galactic source. This discovery inspired the construction of the first radio telescopes, after which things developed rapidly.

---

[4] Microwaves are sometimes considered to be very short radio waves.

In the 1950s, a group of British scientists at Cambridge University undertook a special project whose objective was to map the skies. Multiple sources were revealed and classified, leading to the compilation of the first catalog of radio sources, which is still in use today. In 1967, the same group discovered the first pulsar, a celestial object that will be discussed in detail in the third part of this book, a discovery that earned the group leader a Nobel Prize. Radio waves are also used to measure the content of non-ionized hydrogen in galaxies and galaxy clusters, to detect molecules within the galaxy and outside of it, and to study the universe, as was related in Chapter 1, Section 5.

Telescopes used to observe radio waves are based on the same principle that guides the construction of antennas designed to receive satellite transmissions and they usually look like huge satellite dishes. To obtain the best angular resolution possible, a method called *interferometry* is sometimes used, which requires an entire array of such telescopes, positioned at large distances from one another.

One of the oldest and most important interferometry facilities is the Very Large Array (VLA) located in New Mexico, which is shaped like a Y (see Figure 11). Each antenna in the array is the size of a building several stories high. The antennas are installed on tracks so that the distances between them may be changed according to need; the length of the largest baseline possible is about 15 kilometers. The VLA may sometimes be seen from the air when flying over New Mexico. Many important discoveries were made at this facility over the years of its existence, some of which will be described in the following chapters.

The VLA is not, however, the largest radio interferometer in the world. The largest radio interferometer in the world, the Very Long Baseline Array (VLBA), is made up of different antennas that are scattered throughout the globe, so that the baseline of this array (in other words, the distance between the two antennas that are the farthest from one another) is of the order of Earth's radius (see Figure 12).

This array offers an astronomical device with the best resolution in the world, and provides the ability to observe very small regions of distant systems, and specifically to observe matter in the surroundings of giant black holes in near galaxies, as will be described later on. Each of the array's antennas can also be used separately as an independent telescope.

Figure 11.   The Very Large Array (VLA) located in New Mexico, USA. Image Courtesy of NRAO/AUI.

6. Infrared Radiation

Infrared radiation is emitted in the wavelength range between microwaves and visible light. Hot bodies whose temperature is of the order of the average temperature on Earth emit primarily infrared radiation, and can be discerned also at night using appropriate detectors. Indeed, some night vision devices are based on the detection of infrared radiation emitted from bodies that are warmer than their environment (although other methods exist as well). Stars and galaxies emit infrared radiation that enables us to learn about the chemical composition of the matter, about interstellar "dust" (as seen in the galaxy photos presented in Chapter 2), about different radiation mechanisms and more. Telescopes designed to detect infrared radiation are very similar to ordinary, visible light telescopes, but since infrared radiation is strongly absorbed by water vapor in the atmosphere, infrared telescopes are installed on aircraft that travel at extreme

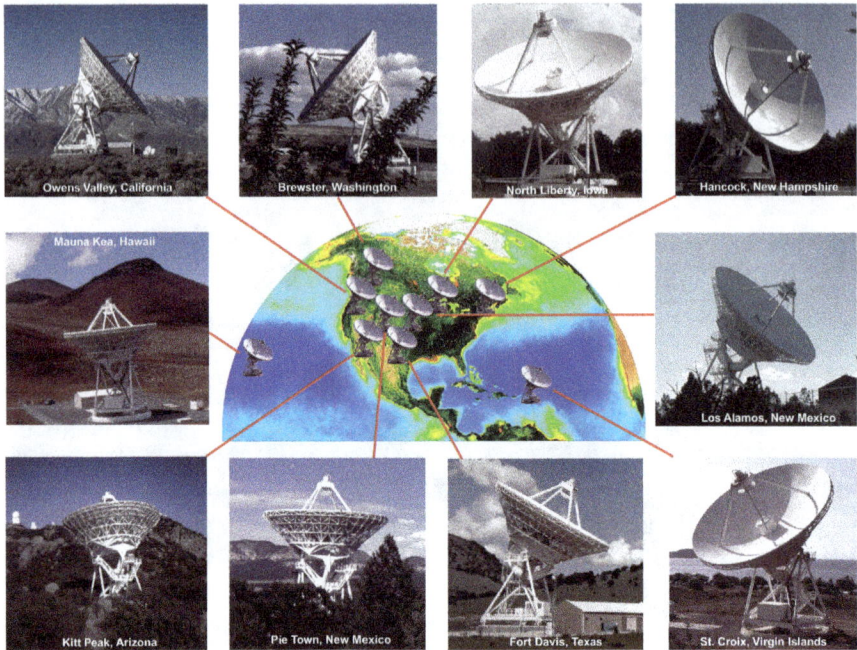

Figure 12. Very Long Baseline Array (VLBA), the largest radio interferometer in the world. The locations of the different antennas are indicated. Image Courtesy of NRAO/ AUI and Earth Image Courtesy of the SeaWifs Project NASA/GSFC and ORBIMAGE.

heights, such as NASA's SOFIA project, or on satellites, like the Spitzer Space Telescope launched in 2003 and the Herschel Space Observatory launched in 2009. One of the most important advantages of infrared observations is the ability to discover galaxies and other objects that are located at the edge of the universe. As we explained in previous chapters, due to the expansion of the universe, light emitted by very distant objects undergoes a redshift, so that a distant galaxy emitting visible light will appear to an observer on Earth as if it is emitting infrared radiation.

7. Visible Light

Observing visible light is the oldest method of astronomical observation. The invention of the telescope is customarily attributed to two Dutch lens makers, Hans Lippershey and Zacharias Janssen, although historians are divided as to whether they are really the first to build such a device. The

What is Interferometry?

Interferometry is a method in which phase differences between waves that are received at several stations located at great distances from one another are measured by means of interference. The technique enables scientists to determine the location of the wave sources with great precision. An array of detectors that measure waves in a coordinated manner is referred to as an interferometer.

The basic concept is demonstrated in the following illustration for a two-station array: a wave is emitted from a very distant source and is received by two interferometer antennas that are at a distance d from one another. The phase difference between the wave received at Station A and the wave received at Station B depends on the wave's angle of incidence relative to the base of the interferometer (in other words, relative to the line that connects the two stations). Measuring the relative phase enables to determine the location of the source in the sky. The angular resolution of the interferometer is equal to the ratio between the wavelength λ of the incident wave and the distance d between the two stations.

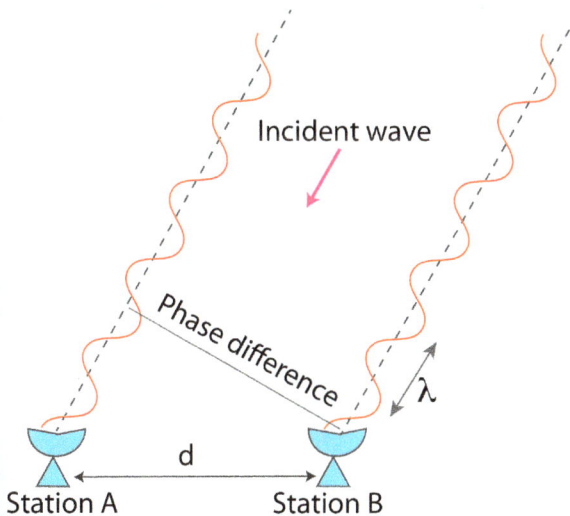

telescope Lippershey constructed in 1608 had a weak magnifying ability, only three- or four-fold. Galileo Galilei, who heard about the invention, significantly improved the quality of the telescope's lens, as well as its structure. The telescopes Galileo built in 1609 were a meter and a half long, had a 20-fold magnifying power, and a much better image quality than Lippershey's instrument. Galileo was also the first to use a telescope for astronomical observations, through which he discovered, among other things, the scarred face of the moon, Jupiter's moons, and Saturn's rings.

Lippershey's and Galileo's telescopes, and later telescopes developed by Johanes Kepler and others, were all based on lenses, similar to magnifying glasses. As it turned out, such telescopes were riddled with limitations, including smeared images, the appearance of a colorful halo due to diffraction within the lens, absorption of part of the light passing through the lens and more. In addition, since the light must travel through the lens itself, it can be supported only around the circumference, making it very difficult to build large telescopes, since the heavy weight of the lenses causes them to warp. Newton proposed an elegant solution to these problems and, instead of the glass lens, used a concave mirror that reflects the incident light to its focal point. Since light does not pass through the mirror but rather is reflected by it, all issues involving diffraction and absorption are avoided. Another advantage is that the mirror, as opposed to the lens, may be supported from the bottom and so it is possible to build telescopes with large baselines, and in fact, all modern telescopes are mirror-based.

Since the days of Galileo Galilei, telescopes have undergone many modifications, although the basic principle has remained the same. The main challenge has been to manufacture large mirrors that comply with the requirements. Technological advances in the past 30 years enabled the construction of giant telescopes, the largest of which contain mirrors with 10-meter diameters. One of the most famous telescopes, Keck, is located on a Hawaiian mountaintop and belongs to a network of prestigious US universities. Since its inauguration in 1993, this telescope has been involved in some of the most important discoveries ever. Several years later, a twin telescope was added (see Figure 13), enabling scientists to apply the technique of visible-light interferometry.

Additional giant telescopes, similar to Keck or the VLT, which is located in Chile and belongs to the European Southern Observatory, are operated by astronomy institutes in Europe and in other countries.

Figure 13. Keck, the world's largest telescope, is positioned at the top of Hawaii's mountain Mauna Kea. The photo shows a pair of telescopes that together enable astronomers to use interferometry. Each of the domes houses a telescope with a 10-meter diameter mirror (which are visible through the dome openings). Image Courtesy of NASA/JPL.

As mentioned previously, the larger the telescope, the more light it receives and so enables the detection of paler and more distant objects. This is the significant advantage that giant telescopes offer. On a theoretical level, the longer the telescope's baseline, the better the potential angular resolution as well, but because of the continuous atmospheric distortions mentioned earlier, light waves penetrating the atmosphere are randomly shifted (twinkled), limiting the resolution of ground-based telescopes. To overcome this problem, a telescope was launched into space in 1990 (Figure 14). This telescope, named after Edwin Hubble, whom we mentioned in Chapter 1, indeed produced the sharpest images ever obtained and exposed new information on the nature of many astronomical systems.

Another method of overcoming atmospheric distortion, which has been implemented for many years in military applications and only recently has

Figure 14.   The Hubble Space Telescope. Courtesy NASA/STSci.

been introduced for use in astronomy, is adaptive optics. Observatories that implement this method are equipped with a telescope whose mirror is attached to a system of small electrical motors. Adjacent to the telescope is a powerful laser device that lights the sky (Figures 15 and 16). The laser beam is reflected back from the atmosphere, and a special computer provides real-time analysis of the deformations of the reflected beam, which are caused by the atmospheric turbulence. The computer then commands the electrical motors to deform the surface of the mirror accordingly so as to offset the atmospheric distortions, thus significantly enhancing the resolution of the telescope and approaching the maximum possible resolution. Such facilities have only recently become available for broader usage.

Instrumentation associated with telescopes has also undergone additional, significant improvement. Photographic film has been replaced by CCD detectors. (Charged coupled device; CCD detectors are present in all digital photography devices. The development of these detectors enabled

Figure 15. A telescope with an adaptive optics system. The green beam emitted through the open dome is a powerful laser beam. The reflections of this laser beam from the atmosphere are used to map atmospheric distortions in real time and to respectively correct the telescope mirror in order to offset these distortions. Courtesy of Javier Mendez (Isaac Newton Group of Telescope, La Palma).

the transition from film-based cameras to digital cameras and video devices, which are used widely and can be found in almost every household.) Advances were also made in the construction of instruments that measure the electromagnetic spectrum, and it is currently possible to measure the spectra emitted by various sources in the universe with a very high level of precision, enabling scientists to significantly increase their knowledge about the matter that stars and galaxies are made of and the matter emitted in supernovae and gamma ray bursts, as well as to measure their velocity and distance from Earth using the Doppler effect.

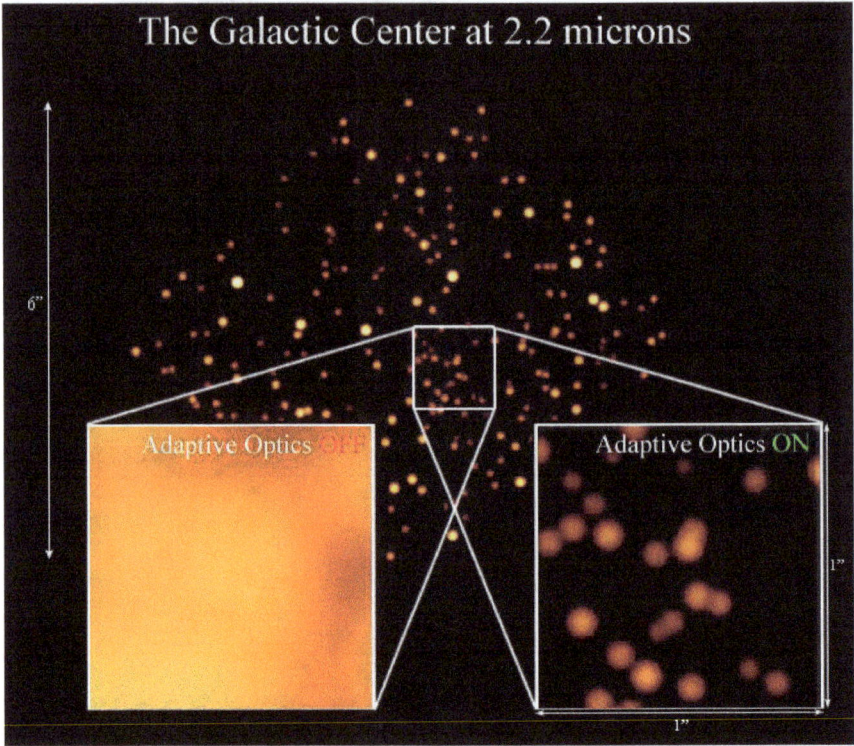

Figure 16. Photos of the center of our galaxy taken with the adaptive optic system off (left) and on (right). The difference in image sharpness is clear. Courtesy of Andrea Ghez, UCLA.

8. X-Rays

Radiation with wavelengths shorter than that of visible light, and especially X-rays and gamma rays, is fully absorbed by the atmosphere, and its detection therefore requires launching detectors into the upper atmosphere or space. Such radiation is lethal for living creatures, and our existence on Earth is possible only thanks to the protection the atmosphere affords against this radiation as well as cosmic radiation, which will be discussed later on. The definition of X-rays is primarily historical. Nowadays, this radiation is customarily defined as electromagnetic radiation with wavelengths ranging between 10 nm and one hundredth of a nanometer. This definition, however, is not the only one: X-rays are often defined in the professional literature according to the given context (the

Figure 17. Wilhelm Röntgen.

kind of equipment used to create the radiation or a certain astrophysical process, for instance). Nevertheless, the overlap between the definitions is usually quite extensive.

Röntgen radiation, which is more commonly known as *X-rays*, is named after Wilhelm Röntgen (Figure 17), who first discovered it — a discovery that earned him the first Nobel Prize in Physics ever awarded.

Röntgen discovered the radiation in 1895, by chance, while studying the emission properties of cathode tubes in his laboratory. At that time, it was known that passing a current through a cathode tube causes a green glow to appear, which is emitted from beams of high-speed electrons and is called *cathode rays* (Figure 18). According to one version, Röntgen covered the tube with a thick piece of black paper in order to darken it out while engaging in another problem. To his surprise, he discovered that a fluorescent screen positioned near the tube began to glow as well, even though the cathode tube was covered. Röntgen suspected that radiation was being emitted from the tube that had more penetrative power than the green glow or any other then-known radiation, and that could penetrate the thick black paper. Since Röntgen had no clue as to the nature of the radiation he discovered, he called it *X-rays*. After the discovery was published and its importance was comprehended, the radiation was named after its discoverer — Röntgen radiation. Both names have stuck and are both still in use today.

In a series of experiments he conducted following this discovery, Röntgen found that the radiation penetrates various materials to various

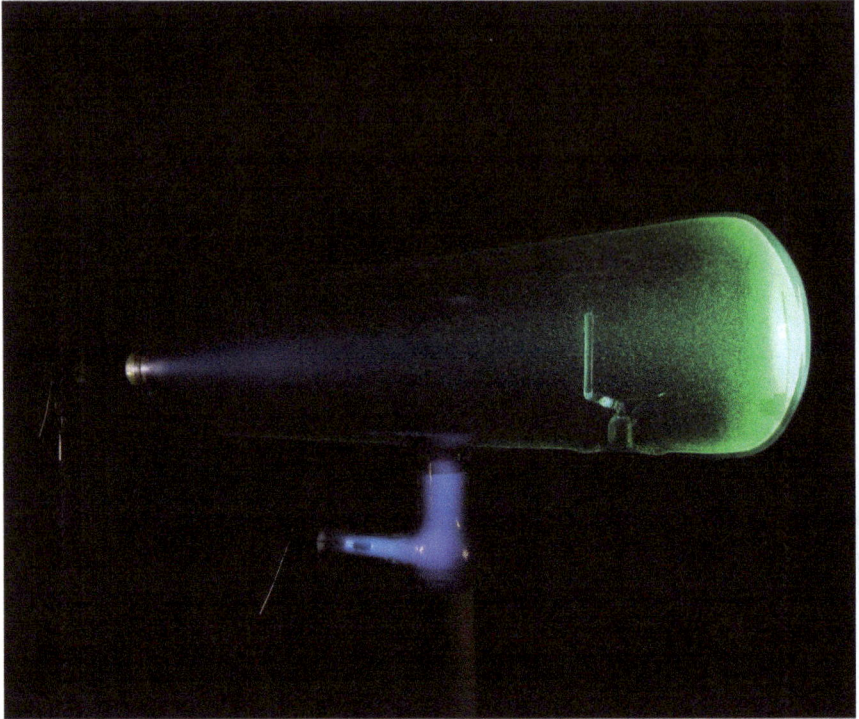

Figure 18. A cathode tube. The greenish glow is caused by electrons that accelerate when the tube is supplied with electrical voltage. Röntgen discovered that when the voltage supplied is high enough, the tube emits X-rays. Courtesy of D-Kuru/Wikimedia Commons.

degrees. Some materials are almost completely transparent to the radiation while other, denser materials, like lead, absorb a major proportion of the radiation. Röntgen also revealed that the radiation penetrates human tissue but is strongly absorbed by the bones, and that this fact can be exploited in order to "photograph" internal body organs. The first X-ray photo ever taken was of Röntgen's wife's hand.

Although X-rays were suspected of being a kind of electromagnetic radiation, this was finally proved in 1912, when German physicist Max von Laue showed, with the help of some colleagues, that when X-rays penetrate a crystal, they are diffracted. Von Laue also succeeded in measuring the wavelength of the radiation, demonstrating that it is significantly shorter than that of visible light and of other kinds of radiation, a fact that accounts for its high penetrability. This discovery also paved the way for the development of crystallography, a technique for determining the structure of crystals from their X-ray diffraction patterns.

X-rays have many applications in a wide variety of areas, including many well-known medical applications; indeed, chest X-rays, dental X-rays, and bone scans in the case of fractures, are only a handful of examples with which we are all familiar. X-rays are often emitted from systems in which particles are accelerated to high energies, like in cathode tubes, synchrotrons (cyclic particle accelerators), or gases that contain heavy ionized atoms, particularly iron atoms.

Many objects in the universe emit X-rays, including the Sun and other stars. As we will see in subsequent chapters, this radiation is emitted by systems in which extreme conditions prevail, such as very high temperatures, strong shock waves, and so on. X-rays and gamma rays, which will be introduced shortly, are therefore important indicators of violent processes taking place in the universe.

X-ray astronomy began developing towards the late 1950s and early 1960s, although the first rocket carrying an X-ray detector was launched into space already in the early 1950s. These initial experiments involved sending rockets and balloons up to heights at which the thinness of the atmosphere enables the radiation to penetrate. Contrary to expectations, the first source of X-rays (not counting the Sun) was found in 1962, in the Scorpio constellation, using a detector mounted on a small, 8-m long rocket. This discovery served as an incentive for additional investments in X-ray astronomy. In 1970, the American space agency, NASA, launched the first research satellite dedicated to X-ray astronomy, which went on to discover nearly 100 additional sources. Since then, various space agencies have sent many X-ray satellites into space, and hundreds of thousands of X-ray sources of different kinds have been revealed throughout the universe. Figure 19 shows the Chandra X-ray Observatory which is still in operation today.

9. Gamma Rays

Gamma rays have higher energy (or shorter wavelengths) than do X-rays. The distinction between X-rays and gamma rays is arbitrary and not agreed upon among scientists. In astronomy, gamma radiation is customarily defined as electromagnetic radiation with wavelengths shorter than

Figure 19. The Chandra X-ray Observatory (named after Nobel laureate Subrahmanyan Chandrasekhar), launched in 1999. Courtesy of NASA/CXC/NGSTC.

one billionth of a millimeter, but this definition is not unanimously agreed upon, and the literature often reveals a certain overlap between the definitions of X-rays and gamma rays.

Gamma radiation is generated during nuclear fission and fusion processes, like those that occur in the Sun or in nuclear reactors; during the nuclear decay of excited atoms; in the annihilation of matter and antimatter; and as a result of the emission by particles that are accelerated to very high energies, similar to X-rays. Many astrophysical systems that emit X-rays also emit gamma rays. Gamma ray observations are conducted primarily using satellites, except in the case of very high-energy gamma rays, where the Earth's atmosphere may be used as a detector. The detection method used depends on the energy of the gamma rays we wish to measure. The technology implemented in astronomy is borrowed from gamma ray detectors used in particle accelerators. This technology has seen dramatic development in the past two decades and is currently based on silicon detectors that provide much better performance than the previous generation of detectors that were based on the ionization of gas. As we will see, gamma-ray astronomy has gained great importance in the study of the universe and its secrets over the past three decades.

In the early 1990s, NASA launched a satellite referred to as the Compton Gamma Ray Observatory (or CGRO for short). This satellite was equipped with four different devices, as seen in Figure 20, for the detection of X-rays and gamma rays of various energies, each of which was designed for a different, unique purpose. For almost a decade after it was launched, CGRO made many sensational discoveries that transformed science's understanding of processes that take place in the universe. In 2002, after completing its mission, the satellite was dropped to the sea in a controlled manner and was given a royal burial.

The many discoveries and plethora of information derived from the measurements performed by the different satellite-mounted telescopes

Figure 20. The Compton Gamma Ray Observatory (CGRO) passing over Earth. The various telescopes mounted on it are clearly visible. Courtesy of NASA.

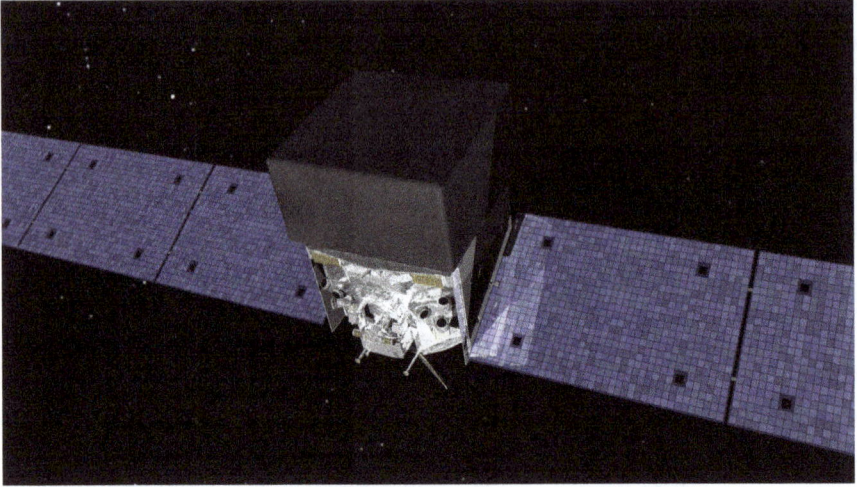

Figure 21. The Fermi satellite is equipped with silicon-based detectors. Courtesy of NASA.

led to the launching of additional satellites by the space agencies of the US, the EU and Japan, implementing the newly acquired understandings and heralding a new age in astronomy. In 2008, NASA launched a new satellite to replace CGRO. The satellite was named Fermi (after the physicist Enrico Fermi) and it implements new technologies based on the use of silicon chips to detect gamma radiation, which were developed originally for use in particle accelerators (Figure 21). The new satellite continues the tradition of discoveries and surprises established by its predecessor. In its very first year in space, Fermi discovered more objects than CGRO did in its entire decade of activity, including several earthshattering surprises, and scientists are anticipating a thrilling future for gamma ray astronomy.

10. The Atmosphere as a Gamma Ray Detector

Very high energy gamma rays are so strongly penetrative that they are impossible to detect using satellites and existing technology. Nevertheless, Earth's atmosphere is still thick enough to absorb this radiation. Thanks to the extremely high gamma ray energies, Earth's atmosphere may be utilized as a giant detector. The technology is based on measuring Chernekov radiation emitted by the products of atmospheric collisions. As mentioned earlier, the wave–particle duality means that radiation may be regarded also as a collection of photons. When a gamma ray photon penetrating Earth's atmosphere encounters the upper atmosphere, it is absorbed and its energy is used to generate a "shower" of electrons and positrons. These particles travel in the atmosphere at a speed that is higher than the speed of light in air (which is 99.97% of the speed of light in vacuum), emitting pulses of blue Cherenkov radiation. When the energy of the gamma photon absorbed in the atmospheric gas is sufficiently high, the particles in the shower emit enough light, cumulatively, so that it can be measured from Earth (see Figure 22). Giant mirrors positioned on the ground reflect the pulses of emitted light to a special camera installed at the mirror's focal point. An

Figure 22. Schematic illustration of a Cherenkov telescope array.

analysis of the shape of the light spot and the pulse period enable scientists to reconstruct the direction and energy of the gamma ray photon prior to its absorption. Devices that use this method to detect gamma radiation are called *Imaging Cherenkov Telescopes.*

Several observatories across the world apply Cherenkov imaging to detect gamma rays. Some, like MAGIC on the Canary Island of La Palma, are equipped with a single giant mirror, while others, like VERITAS in Arizona, US and HESS in Namibia, use an array of mirrors (see Figures 23 and 24 for example).

Figure 23. The HESS telescope array uses the Cherenkov method to measure gamma rays. Courtesy of the HESS Collaboration.

Figure 24. MAGIC: An imaging Cherenkov telescope with a single, 17-m diameter mirror. Courtesy of Robert Wagner, Max Planck Institute for Physics, Munich.

Chapter 7

The Elusive Neutrino

Neutrinos are the strangest, most elusive particles of all particles in nature, and at the same time they are among the most important of them. Neutrinos play a key role in nuclear fusion and fission processes, and can edify us regarding the properties of matter, the elementary forces, the structure of stars, the evolution of the universe, and perhaps also, so the scientists hope, about black holes. The accepted symbol for neutrinos in the scientific literature is the Greek letter v and, as its name implies, it is neutral. In other words, neutrinos carry no electrical charge. The idea of the existence of this particle was born as a wild theoretical hypothesis aimed at solving a contradiction between one of the most basic laws of physics — the energy conservation law — and the results of beta-decay experiments conducted in the 1920s. Only in 1956, over a quarter of a century after the concept of the neutrino was first proposed, did a physicist by the name of Frederick Reines succeed in providing empirical evidence of the existence of this particle, work for which he was awarded the Nobel Prize in 1995.

To understand the sequence of events that led up to the discovery of the neutrino, let us go back in time, to the late 19th century. As mentioned earlier, in 1896 French physicist Henri Becquerel discovered radioactivity, which later on was found to be related to the spontaneous decay of unstable atoms (radioactive elements) and their transformation into other atoms. Although the composition of the atomic nucleus was not yet known at the time, various experiments revealed that during the radioactive decay process, electrons (which are completely unrelated to the electrons surrounding it) are emitted from the nucleus. This was called beta radiation (and so the process is sometimes called *beta decay*). Theoretical calculations based on the momentum and energy conservation laws predicted the energy of the electrons emitted in the process, but the measurements did not match the calculations. The experiment was repeated hundreds of times, but the mystery only grew deeper; the results indisputably contradicted the energy conservation law, one of the most basic laws of physics, which up until that time was in no way subject to any controversy at all. The discovery sent the scientific world into turmoil. The leading physicists at the time were rendered helpless and at a loss for an explanation, and there were even some who, in their despair, questioned the validity of the energy conservation law. Finally, Wolfgang Pauli, the same talented scientist who was mentioned earlier with respect to the uncertainty principle, proposed a surprisingly simple solution. He asserted that in addition to the electron, another particle is emitted in the beta decay process, which travels so easily through any material that it is rendered undetectable. This particle is the neutrino.[1] In fact, only after the neutron[2] was discovered in 1932 did the nature of the beta decay process become clear. It turned out that during radioactive decay, one of the neutrons in the nucleus transmutes into a proton. Since the neutron has no electric charge, the law of electric charge conservation requires that another particle with a negative charge be emitted; this particle is the electron that was detected in the experiments. And in order to satisfy the energy and momentum conservation laws, a neutrino is also emitted, as Pauli suggested. The transmutation of one of the neutrons into a proton is what changes the atom (see Figure 1). This is the basis of every nuclear fission or fusion process that occurs in nuclear reactors, atom bombs, or in the core of the Sun and other stars (thanks to which life on Earth is possible).

---

[1] It is now known that during the beta decay process an antineutrino is in fact emitted, which is the corresponding antiparticle to the neutrino. In some radioactive decay processes, a positron and neutrino are emitted, rather than electron and antineutrino.

[2] The reason for the similar names is historical. Pauli originally called his particle a neutron, since it lacked electrical charge. Shortly afterwards, when James Chadwick discovered a new chargeless particle, he inadvertently also called it a neutron. To avoid confusion, Enrico Fermi, who developed the beta-decay theory, suggested that the smaller particle be called a neutrino (which in Italian means a small neutron).

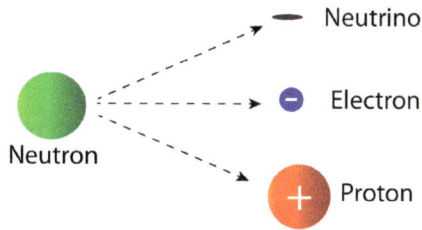

Figure 1. Beta decay.

1. Three Flavors of Neutrinos and Three Generations of Leptons

In Chapter 4 we mentioned the existence of a dozen elementary particles of matter: six quarks and six leptons. The lepton family consists of three electrically charged particles — the electron, the muon, and the tau particle — as well as three chargeless neutrino particles. This division is not random and a connection exists between the neutrinos and the charged leptons; each electrically charged lepton corresponds to a neutrino of the same "flavor". An electron neutrino (in other words, an electron-flavored neutrino) corresponds to an electron, a muon neutrino corresponds to a muon, and a tau neutrino corresponds to a tau particle.[3] Each such pair defines a "generation" in the standard model that unifies the forces of nature, where generations are distinguished from one another by the mass of the charged particle in each pair. The particle generations may be likened to the three bears in the story of *Goldilocks and the Three Bears*: the tau particle, which is the heaviest of the three, resembles Papa Bear, the muon is like Mama Bear, and the electron, which is the lightest, is like Baby Bear. And like the bears' three bowls of porridge, so are there three kinds of neutrinos. The tau neutrino represents Papa Bear's "too-hot" porridge, the muon neutrino represents Mama Bear's "too-cold" porridge, and the electron neutrino represents Baby Bear's "just-right" porridge; each bear and its corresponding porridge represent one generation. The

---

[3] According to the elementary particle theory, all six leptons are related to each other through symmetry referred to as "flavor". Similarly, the six quarks are also related to each other through "flavor" symmetry.

neutrino, discovered in the beta decay process described above, is the electron neutrino. The important point to remember from this analogy is that there are three kinds, or flavors, of neutrinos.

Thousands of experiments over the years have confirmed the standard model and shown that indeed three generations exist in nature. Why three and not four or two? No one knows. Like many other things, it is simply a trick that nature plays. And what is even more interesting, as we shall soon see, is that unlike the three bears' porridge, the neutrinos can exchange flavors with each other according to need.

2. Neutrinos in Space

What do neutrinos have to do with astronomy? Well, as we already explained, wherever nuclear processes are taking place, neutrino particles are expected to be generated. And in the universe, such processes occur in several cases. We have already encountered some of these processes, but we will mention them again. First, the cores of stars, such as the Sun, function as a small nuclear reactor that, on the one hand, provides the pressure required to support the star against gravity, so that it does not collapse, and on the other hand, supplies the energy that is emitted from the star. Neutrino particles created in the stellar core, while it is active, easily penetrate the star's matter and are emitted into the surrounding space. Second, when such a star exhausts its nuclear fuel, its core collapses and the star undergoes a supernova explosion during which it emits a huge burst of neutrinos in a pulse that lasts for several seconds. Third, in the ancient universe, when the temperature of matter was very high, nuclear reactions, during which neutrino particles were emitted, took place at an elevated rate. In fact, the entire universe is a kind of bath filled with cosmic neutrinos, and the dark matter in the universe may actually be made up of those same neutrino particles. In addition, many scientists believe that high-energy neutrinos are created also in the vicinity of black holes and neutron stars, where, so it seems, the greatest particle accelerators in the universe are in action.

3. Flavor Oscillations and the Solar Neutrino Problem

Once astrophysicists comprehended that the source of the Sun's energy is its inner nuclear core, it became clear that the detection of neutrinos emitted by the Sun would provide a "smoking gun" for the theory of stellar structure and evolution. Detailed calculations accurately predicted the flux of neutrinos that was supposed to penetrate Earth, and efforts were made accordingly to construct a suitable and sufficiently large detector to measure them. The first experiment was designed by experimental physicist Raymond Davis, with the help of theoretician John Bahcall who later served as head of the astrophysics group at the Princeton Institute for Advanced Studies. The experiment was set up inside South Dakota's Homestake gold mine, and was operating from 1970 until its end in 1994. A huge tank containing 380 m^3 of a chemical cleaning solution served as a detector, enabling the "capture" of some of the electron neutrinos traveling through the detector (the Sun emits mainly electron-flavored neutrinos).

Very quickly, it became evident that there was a problem. Neutrino particles were indeed discovered, but only about one third of the quantity predicted by the theory. The other two thirds had disappeared. At first, scientists thought something was at fault with the experiment or with Bahcall's theoretical calculations, but after repeating the measurements over several years, paying meticulous attention to the most minute details, and after other people reviewed the calculations as well, there was no escaping the conclusion that only about a third of the neutrino flux reached the detector. Additional neutrino detectors of various kinds were later constructed in different locations throughout the world, and they all supported Davis and Bahcall's conclusion. Despite this evidence, a heated debate ensued among scientists as to the meaning of the measurements, and the phenomenon was termed the *solar neutrino problem*. In 1998, the cause of the phenomenon was gradually clarified following a Japanese experiment, which will be described immediately, and shortly after that, in 2001, decisive evidence was provided of the reason for the shortage of neutrinos from the Sun, following a Canadian experiment designed especially to tackle that issue.

It was found that neutrinos can change their flavor while traveling through vacuum, after covering a sufficient distance. For instance, on its way from the Sun to Earth, an electron neutrino can transform into a muon neutrino and back into an electron neutrino, and so on and so forth. The other neutrino flavors undergo such transformations as well. This phenomenon reflects yet another trick played by the quantum theory, which once again proves to us just how complex natural systems are. The important point is that even if neutrinos of only one flavor are emitted from a given source (like the Sun), they will change flavors while in motion, and if the distance they travel is long enough, a mixture of all three flavors will eventually result and an equal number of neutrinos of each of the different flavors will be obtained (socialism at its best!). Since Davis' original experiment could only measure the electron neutrino, the value obtained corresponded to only one third of the calculated flux — the other two thirds turned into muon neutrinos and tau neutrinos on the journey from the Sun to Earth and, therefore, were not measured by the detector. The Canadian experiment mentioned above was designed especially to measure all three flavors and it indeed revealed that all of the neutrinos emitted by the Sun do reach the detector according to the predictions, only not in their original flavor.

This was one of the rare occasions in which what started out as an astronomical measurement problem led to an important scientific breakthrough — in this case in the area of particle physics.

4. Neutrino Telescopes

Although neutrinos are everywhere in the universe, their detection is a challenge since only a minute amount of neutrinos emitted actually end up striking matter. This explains why experiments designed to detect neutrinos require large volumes containing thousands of tons of detection material. Such experiments are usually located in deep mines to prevent contact with cosmic rays, which creates effects in the detector that are similar to those created by neutrinos, making measurements even more difficult. Several experiments, which are based on different technologies and are designed to detect neutrinos from the Sun and from close supernovae, are currently being conducted in various locations worldwide. We already mentioned Davis and Bahcall's historical experiment in which they used a cleaning agent as a detector, and we also mentioned the Canadian experiment. Another famous neutrino observatory is the Japanese Kamiokande Observatory. This huge detector, which is located in the Kamioka mine near the Japanese city of Hida, implements the Cherenkov method. The detector is in fact a giant water tank (see Figure 2) covered by a dome mounted with thousands of light detectors designed to measure the

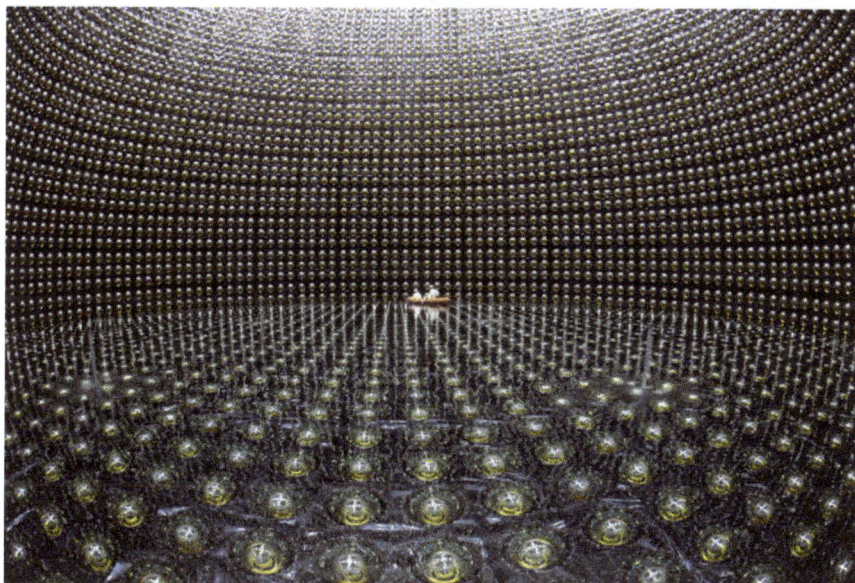

Figure 2. The Super-Kamiokande neutrino detector is situated in the Kamioka mine in Japan. The photo shows the dome containing thousands of light detectors and their reflection in the water-filled tank. On the far side is a boat with two scientists who are engaged in maintenance of the detectors. Courtesy of Kamioka Observatory, ICRR, The University of Tokyo.

Cherenkov radiation emitted by the water. A neutrino particle that strikes the water in the tank knocks out an electron, which then begins to travel through the water at a speed approaching the speed of light. As the electron moves through the water, it emits Cherenkov radiation, which is measured by the photomultiplier tubes in the dome. These measurements make it possible to determine the energy and direction of the incident neutrino.

The Kamiokande Observatory was involved in several important discoveries, in addition to its involvement in the solar neutrino problem (Figure 2). These discoveries include, among others, the first and only neutrino burst from a supernova ever detected, which will be addressed in Chapter 17, and the first measurement in 1998 of the neutrino oscillations discussed earlier. In 2002, the observatory's director, Masatoshi Koshiba, and Raymond Davis were co-recipients of the Nobel Prize, awarded for their important contribution to the study of neutrino astrophysics.

5. What Do Ice Cubes Have to Do with Neutrinos

Another kind of neutrino detectors, currently under construction, are designed to detect neutrinos with energies that considerably exceed those of neutrinos emitted by the Sun or by supernovae. Such high-energy neutrinos may be created when energetic protons accelerated in the vicinity of black holes and neutron stars collide with matter and radiation located near the source. High-energy neutrinos are also formed in the Earth's atmosphere as a result of collisions between cosmic rays and atmospheric gas. Yet another possible source is the decay of hypothetical particles, whose existence is predicted by advanced (and so far speculative) physics theories and which some scientists believe may constitute the dark matter in the universe. The discovery of neutrinos of such high energy, and the identification of their sources, will constitute a breakthrough in the study of astrophysical systems under extreme conditions and will shed new light on some of the processes that take place in those systems.

A neutrino telescope of this kind, whose construction was recently completed, is situated at the South Pole, Antarctica (see Figure 3). Detection is based on the identification of charged particles created when neutrinos collide with ice and rock in the region that contains the detector units. Here too, like in the Super-Kamiokande observatory, Cherenkov radiation is used to identify the charged particles created in the collision and to reconstruct their path. These measurements enable scientists to deduce the original direction of the neutrino colliding with the ice, and its energy. The detector is in fact a kind of ice cube with an overall volume of 1 cu. km, hence its name, the IceCube Neutrino Observatory. Using a specially developed hot-water drill, 86 holes were drilled in the detector region over an area of 1 sq. km. A cable equipped with optical sensors was inserted into each hole, which extends to a depth of approximately 3 km below the surface of the ice. Since tiny air bubbles in the upper layers of the ice scatter the Cherenkov light and interfere with the measuring process, the detectors were located at depths between 1,450 and 2,450 m.

(a)

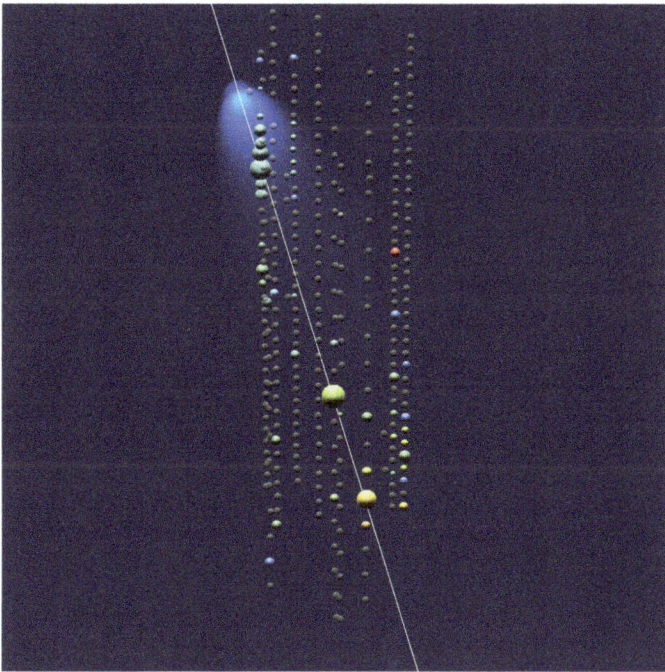

(b)

Figure 3. The IceCube experiment at the South Pole, Antarctica: (a) shows the drilling station. The detector itself is located about a kilometer and a half beneath the surface of the ice, as depicted in (b). The detector consists of thousands of sensors assembled on long, 1-km cables. The slanted white line represents the path of the charged particles created when a neutrino strikes the ice. The bluish light beam on the top left of (b) is the Cherenkov radiation emitted by the newly created particles. Courtesy of NSF/IceCube Collaboration.

The construction of the detector began in 2005 and lasted five years. The total cost of the project was $270 million, with most of the funding coming from the US National Science Foundation and the rest from several European countries. The experiment involved the participation of scientists from 30 countries. The first detection of neutrinos that likely originate from outside the solar system was announced in 2013. Since then the IceCube team has reported the detection of additional events.

A similar detector that utilizes sea water instead of ice was constructed in Europe through an Italian-French joint venture. This new detector is located in the Mediterranean, several kilometers beneath sea level.

Chapter 8

Gravitational Waves: Spacetime Vibrations

So far, we have addressed electromagnetic radiation and its discovery as well as the properties of neutrino particles. Nature, however, provides us with another, completely different kind of radiation that is related to the gravitational force. One of the most important predictions of the theory of general relativity is the existence of gravitational waves, or gravitational radiation. Such radiation is not possible according to Newton's gravitational theory, which is an essentially static theory. The basic reason that necessitates the existence of this kind of radiation in the relativity theory is the fact that no information can travel faster than the speed of light. To understand the meaning underlying this last sentence, try to imagine the following thought experiment: Suppose two small masses are located at a given distance from one another. We hold one of the masses in our hand while the other is attached to a spring scale. Since the force of gravity acts between the two masses, they have a tendency to pull closer to one another, but the reactive forces applied by our hand and by the scale cause the masses to remain in place. The scale will, therefore, give a reading that corresponds to the gravitational force acting between the two masses. Since the gravitational force depends on the distance between the masses, the reading on the scale will change if we change the distance between them. Let us now suppose that we suddenly draw our hand back, increasing the distance between the two masses. What will the scale reading be now? Solving Newton's equations yields an instantaneous change in the scale reading. In other words, the information regarding the movement of the mass that is in our hand was transmitted to the other mass, which is attached to the scale, at an infinitely high speed. And here lies the problem, since according to the principles of the relativity theory, no entity in nature can move faster than light. The equations of general relativity contain this principle within them. Indeed, solving these equations instead of Newton's, one finds that the scale does not in fact react immediately, but only after a time that is equal to the time it takes light to travel the distance between the two masses. In other words, the disturbance we created propagated in space at a finite velocity, like a sound wave that propagates through the air when we clap our hands or like a light wave emitted from a flashlight.

In general, when a gravitational field undergoes change at a certain point in space, it takes time before the gravitational field at a more distant point responds. Like electromagnetic radiation that is emitted when an electric charge accelerates, gravitational radiation is emitted due to the acceleration or change in time of a mass (or energy). For example, the rotation of Earth around the Sun causes the emission of gravitational waves (in this specific case, the wave intensity is very low). From a mathematical perspective, the description of gravitational waves is completely identical to the description of sound waves or electromagnetic waves. Like electromagnetic waves, gravitational waves travel at the speed of light and need no medium in which to travel. In fact, the medium in which gravitational waves travel is spacetime. But unlike electromagnetic radiation, in which the electromagnetic field has two kinds of sources — positive charge and negative charge — the gravitational field has only one source (there is no negative mass). As a result, the basic pattern of a gravitational wave is slightly different than that of an electromagnetic wave, but this detail is of importance mainly for experts dealing with the design of detectors and the analysis of signals, and not for the rest of our story.

From the geometric perspective of the theory of general relativity, a gravitational wave represents ripples in the curvature of spacetime, as the following illustration attempts to demonstrate. This phenomenon is reminiscent of the ripples formed when a pebble is cast into water, where the disturbance creates distortions that propagate along the interface between the fluid and the air. In the case of gravitational radiation, the ripples represent changes in the curvature of spacetime instead of in the height of the water surface. This is why the dimensions of an object present in a region traversed by a gravitational wave will change according to the wave pattern. For instance, if a periodic wave were to travel through our living room, we would discover that the dimensions of our sofa change in a periodic manner — sometimes it would be longer and other times, shorter. In fact, the wave changes not only the dimensions of the object it encounters, but also the entire space and its contents.

This property of gravitational waves — the ability to change the dimensions of any object in the region in which it travels — is the basis for most methods of gravitational wave detection. The main difficulty in constructing detectors lies in the fact that for any realistic source of gravitational waves in the universe, the changes that such waves induce in the vicinity of Earth are so small, that measuring them is undoubtedly one of the toughest technological challenges ever, especially in light of the fact that similar changes might be created by other phenomena as well, like seismic noise (small earthquakes) or internal vibrations of the device itself.

Direct detection of gravitational waves plainly reveals the dynamic nature of spacetime and constitutes an additional, important confirmation of the correctness of the theory of general relativity (see Figure 1). Furthermore, new information about many astrophysical systems may be

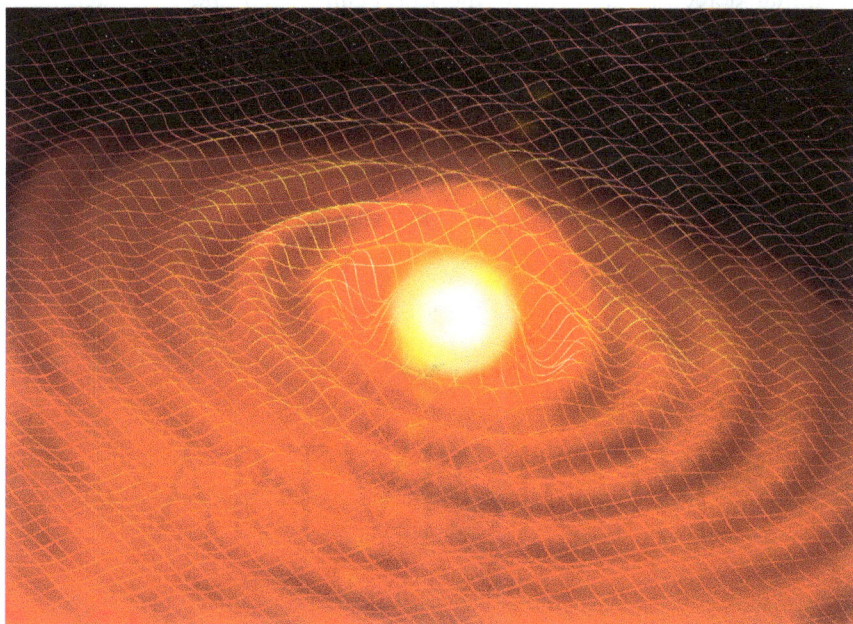

Figure 1. An illustration of a gravitational wave created by the white source in the center. Courtesy of INFN/made by Luca Cesaro.

derived from gravitational radiation measurement, opening a new window to the mysterious universe. Until recently, gravitational radiation has only been measured indirectly, in an experiment that will be described in detail in the third part of this book. The first direct detection of gravitational waves was announced on 11 February 2016, as will be described in the following sections.

Several experiments have been performed over the years in an attempt to detect gravitational radiation, but without much success. A special technique was eventually developed based on laser beam interference, with sufficiently high sensitivity to detect objects in the universe. The development of the method took about three decades, during which a prototype was initially built. The construction of the detector itself was completed in 2002 and is called LIGO, which stands for Laser Interferometer Gravitational Wave Observatory. It was developed and built by Caltech and MIT, with funding from the National Science Foundation, and is located in the US (see Figure 2). An advanced version (called Advanced LIGO) began operating a few years ago, and is still being upgraded. A similar detector was built by the European Union and is located near Pisa in Italy.

Figure 2.   The LIGO gravitational wave detector at Hanford Observatory in Washington, US. Courtesy of Caltech/MIT/LIGO.

1. Powerful Lasers for the Detection of Gravitational Radiation

The LIGO detector depicted in Figure 2 consists of two perpendicular arms, each 4 km long. A laser beam, originating from a laser gun and passing through a beam splitter located at the junction between the arms, is split into two beams that travel within each arm. Both the laser gun and the detector are also housed within the large structure at the junction. Each of the two beams is reflected back to the detector from an ideal mirror positioned at the end of each respective arm, creating an interference image at the point at which the two beams converge. This device is actually a giant Michelson interferometer (see frame below for a detailed explanation). To prevent problems related to the dispersion of the laser beams, the beams travel within vacuum tubes that run the entire length of the arms, making this facility the largest vacuum system in the world. The principle of operation is as follows: when a gravitational wave passes through the detector, it causes a relative change in the length of the arms and, as a result, a change in the phase between the two laser beams that are reflected back from the end mirrors. The phase change creates an interference pattern on the detector, which allows scientists to determine the movement of the arms, at any time, relative to their initial state. This information enables researchers to infer the properties of the gravitational wave that affected the detector.

In fact, the LIGO observatory consists of two identical interferometers, one located in Livingston, Louisiana and the other one 3,000 km (over 1,860 miles) away, in Hanford, Washington. The reason for having two so widely separated detectors is the interferometers' sensitivity to background noises, like earthquakes, various sources of acoustic noise, and so on, that can mimic the desired gravitational wave signal. Two nearby detectors would "feel" the same noise. On the other hand, each of the two widely separated detectors is exposed to different noise. By comparing data from both sites, vibrations that differ between the sites may be subtracted and only identical signals that occur simultaneously in both detectors are measured.[1] This technique for filtering the noise using two or more distant interferometers is

---

[1] In fact, there is a small time delay, of a few milliseconds, between each signal that corresponds to the time it takes the wave to travel between the detectors.

The Michelson Interferometer

A laser beam strikes a beam splitter. Part of the beam is transmitted (green beam) and part of it is deflected at a straight angle (red beam). The two beams — the transmitted beam and the deflected beam — are reflected, by respective mirrors located at the end of each of the interferometer's arms, back to the beam splitter in the center which reflects them to the detector. The interference pattern of the laser beams in the detector depends on the relative length of the two arms, L_1 and L_2, and this length can be changed by moving the mirrors at the ends of the arms. This device was invented by Albert Abraham Michelson (hence its name) and it was used in the famous Michelson–Morley experiment mentioned at the beginning of Chapter 6.

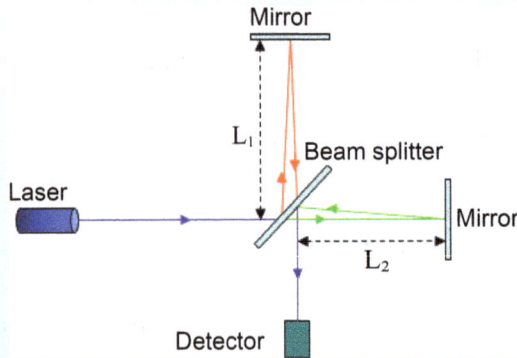

crucial for the detection of gravitational waves. Any detection would be practically impossible using only one LIGO interferometer.

Detectors like the LIGO detector are limited to measuring gravitational waves with a relatively small range of frequencies. At too low frequencies, the effect of ground vibrations (seismic noise) is especially critical and renders it impossible to perform measurements at the accuracy required in order to detect the gravitational waves. At too high frequencies, the sensitivity threshold of the device is determined by quantum phenomena. To understand the huge technological challenge facing the experiment developers, we will mention that operating the interferometer requires remote control of the distance between the mirrors to the extent of a single nanometer (one billionth of a meter). The level of control is achieved using magnets that are attached to the mirrors and which react to the intensity changes of the laser beam upon leaving the splitter, after the two reflected beams are merged together. In addition, in order to isolate the mirrors from external noises, a most sophisticated array of springs and pendulums was constructed on which the mirrors are mounted.

This complex design yields the highest sensitivity in the frequency range between 100 and 2,000 Hz, the optimal range for detecting radiation emitted from sources with masses on the order of the mass of the Sun, and especially from compact binary systems consisting of neutron stars and black holes. One such system, which we will discuss later on, is the *binary pulsar* discovered by two astronomers, Hulse and Taylor. The frequency of the gravitational waves emitted from binary star systems is equal to twice the rotational frequency of the stars around each other. In the case of neutron stars or black holes, the rotational frequency is approximately several hundred revolutions per second, so that the wave frequency falls exactly within the detector's range of maximum sensitivity.

The first direct detection of gravitational waves was made on September 2015, and was announced in a press conference held on 11 February 2016. The signal was detected first by the Livingston detector and only a few milliseconds later by the Hanford detector, as would be expected from theory. This delay in detection corresponds to the time it takes the wave to propagate between the detectors. Detailed analysis revealed that the signal was emitted during the final stages of a binary black hole merger event that took place in a distant galaxy about 1.3 billion years ago (see Figure 3). The masses of the two colliding black holes were estimated to be about 29 and 36 times the mass of the Sun, respectively. This came as a surprise since scientists did not expect stellar black holes with such large masses to form. Two additional detections of black hole coalescence episodes were reported subsequently, implying that these events are more frequent than previously thought.

2. LIGO in Space

Some of the gravitational wave sources in the universe are expected to emit gravitational waves at much lower frequencies than those detectable

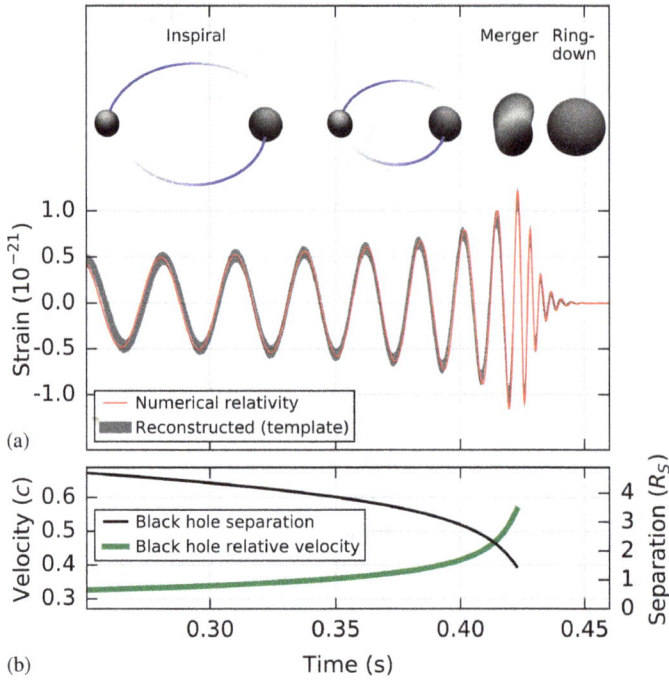

Figure 3.   The first direct detection of gravitational waves from merging black holes by LIGO. (a) The upper panel shows the wave pattern and a schematic illustration of the various stages of evolution. (b) The bottom panel shows the black hole separation and relative velocity. Courtesy of Caltech/MIT/LIGO.

by LIGO. An especially interesting example is a binary system of giant black holes (see Figure 4). In Chapter 2, we noted that the centers of most galaxies in the universe are occupied by giant black holes with masses that range between one million to several billion times the mass of the Sun. Observational evidence indicates that, instead of a single black hole, some of these galaxies contain systems of two giant black holes in close orbit around each another. It is very likely that these systems are the result of the collision between two galaxies, each with one giant black hole in its center. Following the collision, the galaxies merged to create a single new galaxy, whereas their black holes sank to the center due to their large masses. Once the black holes became sufficiently close, they began to revolve around each other.

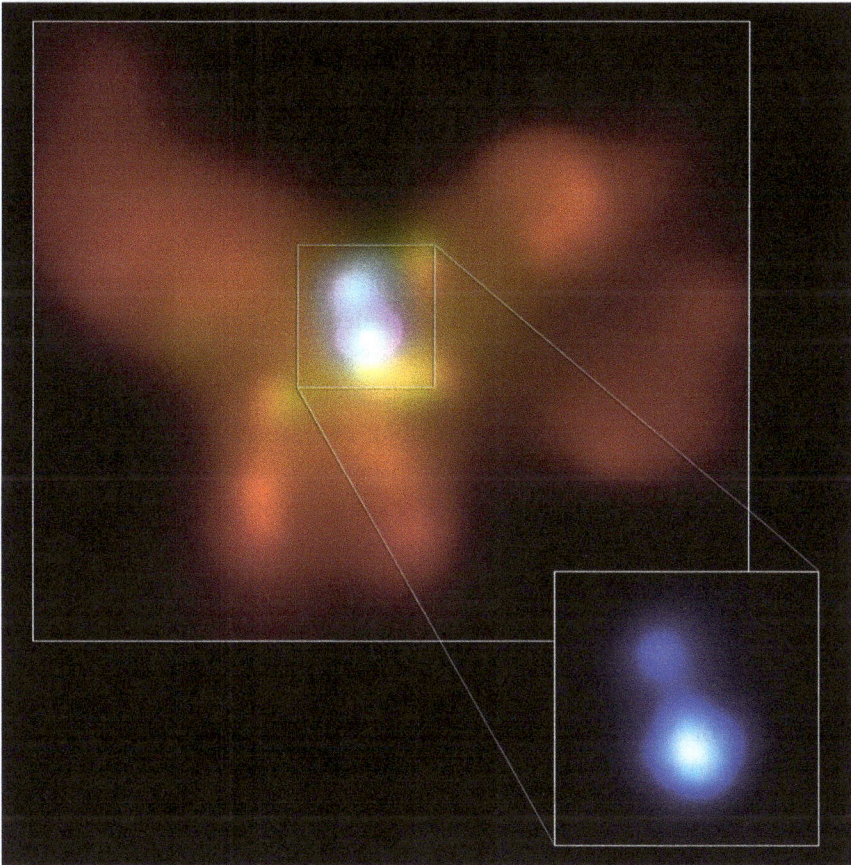

Figure 4. A binary system of giant black holes at the center of the NGC 6240 galaxy, as captured by the X-ray camera onboard the Chandra Space Observatory. Courtesy of NASA/CXC/MPE/S.Komossa *et al.*

Supermassive black hole binary systems are expected to emit gravitational waves, just like the double neutron stars, but due to the system's size, the angular velocity of the giant binary black hole system is much smaller than that of a binary neutron star system (or a binary stellar black hole system), and so the frequency of the gravitational waves they emit are expected to be respectively low. For a typical system in which the black holes are close to collision, gravitational wave frequencies are expected to be between 0.01 and 0.1 Hz. Measuring these gravitational waves will provide an additional, independent way of studying the formation and evolution of galaxies at early cosmic epochs, and of studying the processes that take place at the centers of such galaxies.

Another, albeit speculative, source of low-frequency gravitational waves is primordial fluctuations in the early universe. Different theories predict that during the creation of the universe, extreme processes took place, some of which are connected to the appearance of the forces and others to the possible existence of additional dimensions, which caused strong spacetime vibrations which expanded freely throughout the universe. While the information imprinted in the cosmic background radiation reflects what happened only at the end of the recombination era, when the universe was 380,000 years old, the information carried by the primordial gravitational waves was imprinted in much earlier times, and may possibly edify us as to the processes that occurred immediately after the universe was created. Another possible source of cosmic gravitational waves is related to the superstring theory. This theory predicts the existence of cosmic strings that extend to astronomical dimensions due to the expansion of the universe. Like guitar strings, these strings also vibrate, emitting gravitational waves. The search for these gravitational waves will act as a test of the cosmic strings theory.

In order to optimize signal reception by the interferometer, the length of its arms must equal one half of the wavelength being measured. A gravitational wave with a frequency of 0.1 Hz has a wavelength of 3,000,000 km — 500 times the radius of Earth. Hence, it is clear that the detectors mounted on Earth are ineffective when it comes to measuring low-frequency gravitational waves. In addition, the detector must be isolated from seismic noise, which at these frequencies are especially strong.

To overcome these obstacles, a design was proposed for the construction of a detector based on laser interferometry, similar to the LIGO, but with a much larger baseline. This system, which was named LISA (Laser Interferometer Space Antenna), is destined to be placed in space, in an orbit around the Sun, and to include an array of three satellites positioned at a distance of about 5,000,000 km from one another (see Figure 5). The main satellite will be equipped with a huge laser gun, which will direct laser beams toward the two other satellites. The beams will be reflected back by mirrors that will be installed on the secondary satellites, so that one beam will travel clockwise in a closed path and the second beam with

Figure 5. A schematic description of the LISA system for detection of gravitational waves. Courtesy of ESA.

travel in the opposite direction. The beams will meet up again at the main satellite where the interference pattern will be measured. The passage of gravitational waves will cause periodic changes in the distances between the satellites, which will be measured by measuring the change in the interference pattern of the laser beams traveling between the satellites. In order to measure the minute changes caused by a gravitational wave, the distance between the satellites must be known with almost absolute accuracy. This requires the development of innovative satellite synchronizing technologies, which is currently underway. The LISA project is a joint venture between NASA and the European Space Agency (ESA).

Chapter 9

Cosmic Rays — A Shower of Energetic Particles from the Universe

Cosmic rays, as distinguished from cosmic background radiation, is the name given to a collection of very high-energy particles that reaches Earth from the expanses of the universe. Cosmic rays contain mainly hydrogen nuclei (protons) and some helium, as well as heavier nuclei. In addition to all of these particles, cosmic rays also contain electrons and a minute amount of antimatter. The slightly misleading term, cosmic rays, was coined in the 1920s by Robert Millikan who believed that this radiation was a kind of gamma rays arriving from outer space. Only later did the true nature of this cosmic radiation become clear, but for some reason, the name stuck and is still in use today.

As early as the beginning of the 20th century, and maybe even before, scientists knew that the atmosphere was ionized by some kind of radiation (see Box 1). The prevailing belief was that the source of the ionizing radiation was the radioactive decay of matter in the ground. But measurements of the ionization level at various heights in the atmosphere showed that, contrary to expectations, the intensity of the ionizing radiation increases with the height above the surface of Earth. Parallel measurements revealed a low level of radioactivity in the sea, under water. The conclusion from this entire set of measurements was that the source of ionizing radiation is not subterranean but rather is located in the atmosphere itself or outside of the atmosphere. One hypothesis was that the source is somehow related to the Sun. In a series of measurements conducted by Victor Hess in 1912 using hot-air balloons that floated several kilometers above the ground, he

Box 1.   What is Ionization?

Ionization is a process in which a neutral atom becomes an ion (a charged atom) as the result of the removal or addition of electrons. Atoms may be ionized in a variety of ways. One way is by exposing them to electromagnetic radiation of very short wavelength (ultraviolet radiation and X-rays, for example) or bombarding them with beams of high-energy particles (beta rays, for example). When a particle of ionizing radiation hits an electron in an atom, it knocks the electron out of the atom and leaves behind a charged atom.

determined that the ionization rate of the atmosphere undoubtedly increases with the height. Measurements that he conducted during a solar eclipse, when the Moon fully conceals the Sun, also showed a high ionization rate, and thus the idea that the sole source is the Sun was rejected. This work earned Hess a Nobel Prize. As mentioned earlier, Robert Millikan was the one who proposed that the source of ionizing radiation is cosmic, meaning that it arrives from outer space. Many additional experiments conducted during the first half of the 20th century, alongside the laboratory-based discovery of elementary particles, eventually revealed the properties of cosmic radiation as we know them today.

When a cosmic ray particle collides with an air atom, it knocks electrons out of the atom and ionizes it. This is the source of atmospheric ionization. The deeper the particle penetrates into the atmosphere, the greater the chance it will collide with air atoms and loses its energy, and so, as Hess discovered, the closer to Earth, the lower the ionization rate (since fewer particles reach there). The ionization of the atmosphere by radiation has implications for Earth's climate.

The energy of cosmic ray particles extends over a very broad range, specifically over about 10 orders of magnitude. The most energetic particles have energies that are much higher than the energies that can be obtained in the world's largest particle accelerators. For the sake of comparison, the energy of an ultra-high-energy proton is equivalent approximately to the energy of a tennis ball traveling at a velocity of 100 kph. Lower-energy

Table 1.   Sources of cosmic rays of various energies.

Energy	Source	Notes
Low	Sun and other galactic sources	
Medium	Supernovae in our galaxy	The particles are trapped in the galaxy long after their creation.
High	Unknowns	Apparently of extra-galactic origin.
Ultra-high	Gamma ray bursts? Quasars? Magnetars?	Arrive from outside the galaxy, from a distance that does not exceed 100 million light years.

cosmic rays are apparently accelerated in shock waves created by supernovae in our galaxy. The Sun also emits cosmic rays, but only at very low energies. The energies of the cosmic ray particles emitted from the Sun are 50 billion times less than those of particles with the highest energies ever measured. The origin of the highest energy cosmic rays is not yet clear, but considerable evidence indicates that they arrive from the near universe outside of the Milky Way galaxy. Possible sources of *ultra-high-energy cosmic rays*, as they are called, will be addressed in the third part of the book. The flux of the ultra-high-energy cosmic rays is very low; on average, only one particle strikes 1 sq km of the atmosphere once every 100 years. This is why cosmic ray detectors must have the largest surface area possible. The sources of cosmic rays of various energies are given in Table 1.

1. The Cosmic Ray Horizon and Cosmic Background Radiation

Cosmic background radiation was first mentioned in Chapter 1. It is electromagnetic radiation in the microwave range, a leftover from the Big Bang. This radiation bath, which fills the entire universe, holds important information about the state of the universe during its initial stages of evolution. Every object in the universe — galaxies, stars and so on — is immersed in the cosmic background radiation and feels it. Even cosmic ray particles feel it.

When a particle emitted by a distant object travels through the expanses of the universe, it rubs up against the cosmic background radiation and, as a result, loses energy.[1] The higher the particle's energy is, the shorter is the distance over which it loses a significant part of its energy in collisions of this kind. For an ultra-high-energy particle, this distance is very small compared with the size of the universe: only 100 million light years (about one hundredth of the size of the visible universe). Therefore, ultra-high-energy cosmic rays that enter Earth's atmosphere cannot have been created in very distant regions of the universe, for if so, they would have lost all of their energy on the way. In other words, there is a "horizon" of cosmic ray sources, and it is very close to us compared with the size of the entire universe. This fact is a most important clue as to the nature of these sources and the ways in which they should be sought.

2. Detection Methods

Similar to the detection of very-high-energy gamma rays, the Earth's atmosphere may be used to detect high-energy cosmic rays. A cosmic ray particle (such as a proton or helium nucleus) penetrating the upper atmosphere, collides with air atoms and creates, through a chain process, a secondary "shower" of particles that travel in a narrow cone around the original direction of the particle. The more energetic the penetrating particle, the more secondary particles will be created in the shower. Measuring the parameters of the shower of secondary particles provides the information required to calculate the energy and direction of the initial cosmic ray particle that created the shower. There are two main methods for measuring showers. The first is to measure the shower particles themselves using detectors installed on the Earth's surface. The second method takes advantage of the fact that shower particles heat the atmospheric air, exciting atoms as a result (see Figure 1). Upon cooling, these excited

---

[1] More precisely, the particle occasionally collides with a photon of this background radiation. The particle nature of the radiation is manifested in this process.

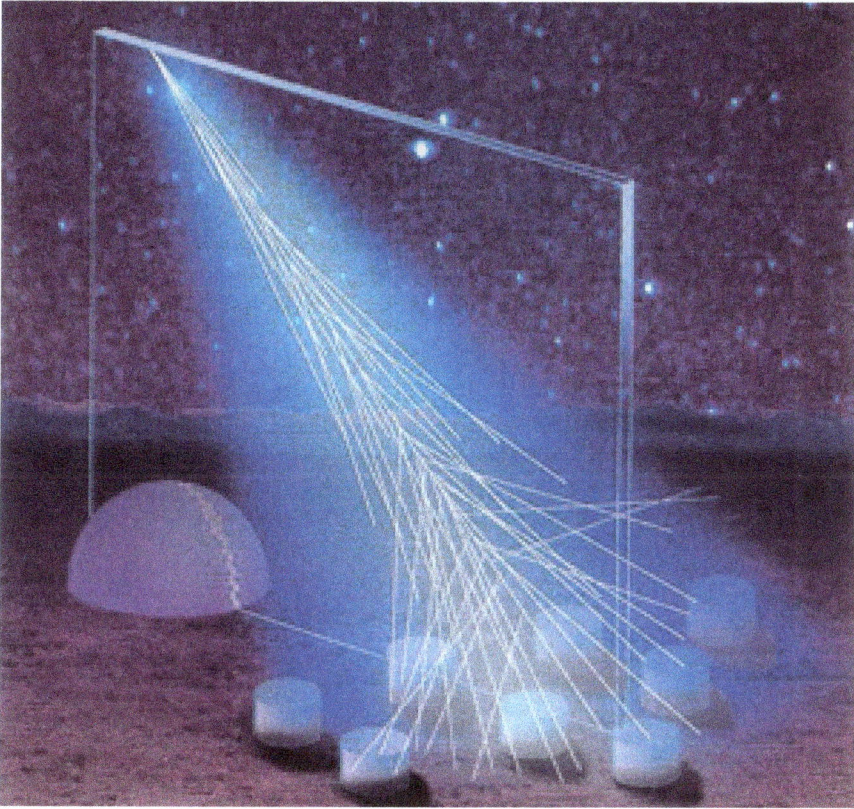

Figure 1. A schematic description of a particle shower created by a cosmic ray that penetrates the atmosphere. The shower particles may be detected directly using detectors located on the surface of Earth or by measuring fluorescent light emitted by nitrogen atoms excited by the shower. Courtesy of Pour La Science.

atoms emit light (a process called *fluorescence*, which is similar to what happens in fluorescent lamps, in which atoms are excited by an electrical pulse), and this light can be measured using mirrors positioned on the ground. Facilities currently in use measure the fluorescent radiation of nitrogen atoms. The Fly's Eye Observatory, located in western Utah implements this method (see Figure 2).

Figure 2. The Fly's Eye Observatory: A pair of mirrors designed to detect fluorescent radiation from showers of cosmic rays. The pair of mirrors enables very accurate stereoscopic measuring of the shower. Courtesy of Christopher Wilkinson of South Australia.

3. The Pierre Auger Cosmic Ray Observatory

A new cosmic ray observatory, which combines the two detection methods, has begun operating in recent years (see Figures 3 and 4). The observatory, which was named after Pierre Victor Auger, a French physicist and well-known researcher in the fields of atomic physics, nuclear physics, and cosmic rays, is located in western Argentina, and is operated by a partnership of 17 countries, including the US, Brazil, Argentina, several European counties, and several East Asian countries. The observatory extends over an area equal roughly to the size of Luxembourg, and it contains approximately 1,600 containers for the direct detection of the shower particles. The distance between each pair of adjacent containers is 1.5 km (which is a little less than one mile). In addition, several units are scattered among the containers to measure the fluorescent radiation of nitrogen. The combination of these two methods in a single experiment offers much more accurate measurements than are possible using only one of the two methods.

Figure 3. The container in the photo contains 12 tons of pure water. When a charged particle hits the water, it emits Cherenkov radiation, which is absorbed by the photo multipliers located within the container. The photomultipliers translate the light intensity absorbed into electric currents. The intensity of light that the particle emits in the Cherenkov process is proportional to its energy. Approximately 1,600 such containers are scattered throughout the area, at intervals of 1.5 km. Courtesy of the Pierre Auger Observatory.

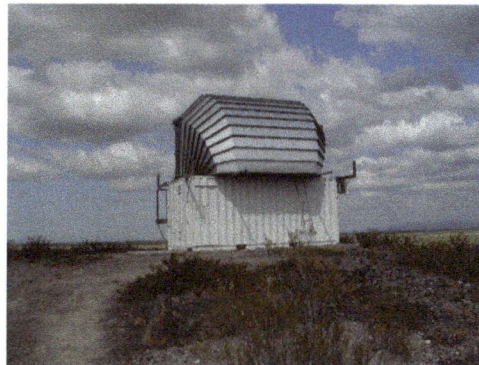

(a)   (b)

Figure 4. A unit for the detection of the fluorescence radiation of nitrogen. (a) shows the detector's mirror; (b) shows the entire structure. The experiment includes several such units that are scattered in the open areas between the containers. Courtesy of the Pierre Auger Observatory.

Third Episode

A Menagerie of Extreme Phenomena

Compact objects are a class of astronomical objects that are much denser than ordinary matter and that play a key role in a variety of extreme cosmic phenomena. The strangest members in this class are black holes, in which the gravitational force is so strong that nothing can escape from inside them. Other objects are neutron stars and white dwarfs, in which mysterious forces, governed by quantum effects, create pressure that offsets the gravitational force and prevents them from collapsing into a black hole. How are they created? How big are they? What is the mysterious quantum force that prevents collapse? How do we know they exist, and what means can be used to observe and identify them? How are they related to the plethora of fascinating phenomena in the universe? These are the questions we address in this final part of the book.

Chapter 10

White Dwarfs and the Quantum Theory

We have seen that a white dwarf is created when a star, whose mass is smaller than about 10 times the mass of the Sun, dies after exhausting the nuclear fuel in its core. White dwarfs are a million times denser than the average density of the Sun (which is approximately equal to that of Earth), so that even though their mass is similar to that of the Sun, their radius is only just one hundredth the radius of the Sun. If the Sun were the size of a tennis ball, a white dwarf will be smaller than a pinhead (see Figure 1), hence its name *dwarf*. Due to their small dimensions and high density, the gravitational force on the face of white dwarfs is a million times stronger than the force of gravity on Earth. Their color, white, results from their high surface temperature, which is caused by the considerable heating of the core during its contraction. Over time, the dwarf cools down and grows pale. Most white dwarfs consist mainly of carbon and oxygen; some, those that were created from especially low-mass stars, are made of helium.

In an ordinary star, like the Sun, pressure is created as a result of the random motion of atoms. The higher the temperature of the matter, the faster the movement of its atoms, and the greater the pressure. This kind of pressure is called *thermal pressure* (that is, pressure that is created by heat). In addition to its dependence on temperature, thermal pressure also depends on the density of the matter. When the matter is completely cold, the pressure is equal to zero. As explained above, the action of the nuclear reactor in ordinary stars causes constant heating of the core, and thermal pressure is therefore maintained as long as the reactor is active. This pressure offsets the gravitational force and prevents the star from collapsing. White dwarfs, on the other hand, have no source of nuclear fuel, and so the colder the dwarf becomes, the lower its thermal pressure. Ultimately,

Figure 1. The bright object in the center is Sirius A, the brightest star in our galaxy. The white dot indicated by the red arrow is a white dwarf that revolves around the star. The photo was taken by the Hubble Space Telescope.

the temperature, and hence the thermal pressure, will drop to a value much lower than is required to maintain the white dwarf in a hydrostatic equilibrium. If so, what then offsets the tremendous gravitational force and prevents the dwarf from collapsing? The answer is the quantum pressure, or in professional jargon *electron degeneracy pressure*. What then is quantum pressure? As we shall see immediately, the answer to this question is connected to one of the most important properties of particles in nature.

1. Quantum Pressure and the Existence of White Dwarfs

In Chapter 4, we mentioned the Pauli exclusion principle, and the fact that electrons abide by this principle, according to which only one single electron can be in a given quantum state (that is, in a state of specific energy and angular momentum). We also explained that the fact that two electrons cannot be in the same quantum state actually enables the existence of atoms and molecules, and more generally, enables the existence of voluminous matter. Like with atoms, it is the Pauli exclusion principle that enables the existence of white dwarfs. More specifically, the source of the quantum pressure, which opposes the gravitational force and prevents the collapse of the white dwarf, is related to the Pauli exclusion principle, which applies to the electrons in the white dwarf's dense matter. However, unlike ordinary matter, in which the electromagnetic force also plays a key role, the size and density of a white dwarf are determined by the balance between its self-gravitational force and the quantum pressure of its electrons. At densities characteristic of white dwarfs, electromagnetic forces are negligible.[1]

When matter is compressed to very high densities, the average distance between the electrons in the atoms decreases. Since the electrons follow the Pauli exclusion principle, they prefer to be as distant as possible from one another, and so they resist the force trying to pull them together. This force, which opposes the compression of the matter, is the source of the quantum pressure. Unlike thermal pressure, quantum pressure depends only on the matter's density, not on its temperature, and it is maintained even when the matter is completely cold. The denser the matter, the higher the quantum pressure. When the core of a star that has exhausted its fuel collapses, the electrons in the collapsing core begin crowding together and the quantum pressure rises as a result. When the density is approximately one million times its initial value, the quantum pressure finally offsets the gravitational force, and the collapse process comes to a standstill. At this point, the collapsed core becomes a small, dense star, or in other words, a white dwarf. It turns out, however, that only objects up to a certain mass can exist as white dwarfs, or more precisely, only objects whose mass does not exceed 1.4 solar masses. When the mass of the contracting core exceeds this value, the gravitational force outweighs the electron degeneracy pressure, and the

---

[1] In fact, the structure of matter in objects whose mass is less than five times the mass of Jupiter, including Earth and most of the planets, is determined by the balance between the quantum force and the electromagnetic force. In heavier objects, the quantum pressure is offset by the gravitational force, and the electromagnetic force plays only a minor, secondary role.

core collapses into a neutron star or a black hole. This critical mass, above which the existence of white dwarfs is not possible, is called the *Chandrasekhar mass*, after Subrahmanyan Chandrasekhar, who was the first to calculate this value.

Chandrasekhar was 20 years old when he made this discovery, while sailing from India, his homeland, to England where he was about to embark on his doctoral studies. At first, his work was scorned upon by British astrophysicist Arthur Eddington, who was one of the most important and influential scientists at the time. Eddington's attitude prompted Chandrasekhar to leave England and immigrate to the United States, where he lived and worked for the rest of his life. His research on the structure and evolution of stars met with great success and publicity, and even earned him the 1983 Nobel Prize. After his death in 1999, the American space agency NASA decided to honor Chandrasekhar by naming the X-ray satellite it launched into space that year after him.

The gravitational force in neutron stars is offset by the quantum pressure of the neutrons, which, like the electrons, follow the Pauli exclusion principle. Since neutrons are heavier than electrons (about 2,000 times heavier), this balance occurs at much higher densities. The ratio between the radius of a neutron star and that of a white dwarf is roughly equal to the electron–neutron mass ratio. In other words, the radius of a neutron star is about 1,000 times smaller than the radius of a white dwarf, while its density is about five billion times greater than that of a white dwarf. As we will explain in Chapter 11, additional effects that exist at such densities enable the existence of neutron stars with masses greater than the Chandrasekhar mass.

2. White Dwarf Tango and Nuclear Explosions

Like most human beings, stars too may be in a relationship. Common wisdom holds that a large fraction of the stars in our galaxy, and in the entire universe, do not like to be single, but rather prefer to be in binary or multiple stellar systems. The reason for this is not entirely clear, but is presumably related to the fact that most stars form in clusters rather than singularly. A binary star is a system that consists of two ordinary stars in orbit around each other. When the time comes for one of the stars to die and become a white dwarf, a new binary system is formed in which one of the stars is an ordinary star and the other is a white dwarf, like in the case of the Sirius stellar system depicted in Figure 1. The gravitational pull of the white dwarf causes a small amount of matter to detach from the ordinary star and slowly attach to the white dwarf (see Figure 2). The accreted matter, mostly hydrogen, accumulates on the surface of the white dwarf, which contains carbon and oxygen, and causes extreme local heating. When the temperature exceeds a certain critical value, approximately 10–20 million degrees, a kind of nuclear explosion occurs on the surface of the white dwarf (a carbon–nitrogen–oxygen reaction; CNO). This thermonuclear runaway reaction, which is also called a *hydrogen flash*, releases a huge amount of energy in the form of heat and radiation, violently ejecting the envelope of burning matter that accumulated on the surface of the white dwarf, like what happens in an atomic explosion, but on a much larger scale. This eruption causes a sudden glow that lasts several days and slowly fades away — a phenomenon called a *nova* (which means *new*). After the explosion, a new accretion cycle begins, leading to another explosion, and so on and so forth.

Novae are not the only phenomenon involving the accretion of matter by white dwarfs. Too high an accretion rate might lead to the explosion of the entire white dwarf in an event called *a Type Ia supernova*, as distinguished from a *Type II supernova*, which is caused by the collapse of a massive star, as will be elaborated on later in Chapter 17.

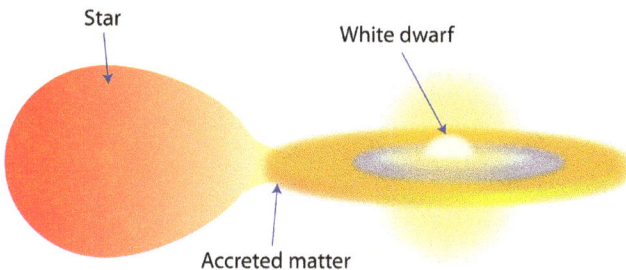

Figure 2. A binary system consisting of a white dwarf and an ordinary star. Matter from the star is transferred to the white dwarf and accumulates on its surface. This accumulation of matter leads to a nuclear explosion, which is seen by the distant observer as a nova.

Chapter 11

Neutron Stars — The Largest Atomic Nuclei in the Universe

As mentioned earlier, objects whose mass is greater than the Chandrasekhar mass cannot exist as white dwarfs and must collapse. If their mass is not too great, their collapse will be halted by the neutron degeneracy pressure, and they will become neutron stars. The existence of neutron stars was predicted by astrophysicists Walter Baade and Fritz Zwicky in 1933 based on Chandrasekhar's work, about one year after the neutron was discovered. Only in 1967, however, did a group of British astronomers discover actual evidence of the existence of neutron stars, in the form of a celestial object referred to as a *pulsar*, a discovery discussed extensively in Chapter 13.

1. The Discovery of the Neutron

Experiments conducted in 1930 with radioactive materials led to the discovery of a very penetrative form of radiation. At first, researchers believed that the radiation consisted of gamma rays[1], although the newly discovered radiation was much more penetrative than the gamma rays they were familiar with. The data could not be explained by the theoretical calculations, and the phenomenon remained an unsolved mystery. In an attempt to get to the heart of the problem, James Chadwick of The University of Cambridge conducted a series of experiments in 1932.

---

[1] The reason for that deduction was that the newly discovered radiation was not influenced by electric fields.

The new findings completely invalidated the possibility that the radiation was gamma rays, and Chadwick consequently proposed that the newly discovered radiation was in fact composed of a collection of uncharged particles with a mass close to the mass of a proton. Seeing that the new particle was electrically neutral, it was called a *neutron.*

Since the mass of a neutron is slightly greater than that of a proton, it follows from Einstein's mass–energy equivalency principle, $E = Mc^2$, that the energy content of a neutron is greater than that of a proton. Since natural systems strive toward a state of minimum energy, free neutrons (that is, neutrons that are not part of an atomic nucleus) tend to decay spontaneously to protons, electrons, and neutrinos, as already mentioned in previous chapters. The mean lifetime of a free neutron, or the time that elapses from its creation until it decays, is about 15 min on average. The opposite process, in which an electron and a proton merge to form a neutron, also takes place but requires the investment of energy (endothermic process) and does not occur spontaneously in ordinary matter, like that which is present on Earth and on the Sun. As we shall see now, this reaction is of utmost importance in the formation process of neutron stars.

We mentioned earlier that the existence over time of stable atomic nuclei is possible only thanks to the presence of neutrons in the nucleus. The strong force, which acts between neutrons and protons, is the "glue" that holds the atomic nucleus together against the electric repulsion of the protons, and prevents it from flying apart. How then does the fact that atoms exist in nature coincide with the instability of neutrons? Well, even though neutrons tend to decay in vacuum, they are stable within the nucleus of ordinary atoms thanks to the same force that unifies the atomic nucleus. In order to turn into a proton, the neutron must overcome this "glue" that binds it to the other nucleons; this already costs too much energy. This may be likened to an attempt to stand a pencil on its point. Even if we succeed in making it stand up, it will not be stable and will fall to a horizontal position on the table almost immediately. But if we glue the tip of the pencil to the table using a strong enough adhesive, it will be able to stand on its tip for a long time thanks to the glue's resistance to the force of gravity that aspires to topple the pencil over. In the atomic nucleus, this glue acts efficiently only when the nucleus does not contain too many neutrons, or more specifically, when the number of

Box 1.   What is an Isotope?

Isotopes are atoms of the same chemical element that differ from one another in the number of neutrons they have in their nuclei (that is, they have different mass numbers). All isotopes of a given element have the same atomic number (that is, the same number of protons). For instance, the most common isotope of hydrogen has one proton in its nucleus. Deuterium is an isotope of hydrogen that has one proton and one neutron, while tritium is another isotope of hydrogen that is generated in nuclear reactions and has one proton and two neutrons.

Since the chemical properties of an element are influenced only by its atomic number, all isotopes of a given chemical element have identical chemical properties. On the other hand, the number of neutrons in the nucleus affects its stability. Some isotopes, especially the most common natural isotopes, are stable and have an unlimited lifetime. Carbon-12 and oxygen-16 are examples of stable isotopes. Unstable isotopes also exist; however, they tend to decay spontaneously into more stable isotopes. These unstable isotopes are called *radioactive isotopes*. Tritium and carbon-14 are examples of radioactive isotopes.

neutrons approximately equals the number of protons[2]. Such atoms are stable. Additional neutrons may be added to the nucleus of such an atom without changing the number of protons — in other words, without changing the kind of atom and its chemical properties — to obtain a new isotope of the same atom (see Box 1).

In isotopes with too many neutrons, the nuclear adhesive is not strong enough so as to stabilize all of the neutrons in the nucleus, and the resulting atom is unstable. Such an atom will decay after a while and turn into another atom due to the transformation of one of its neutrons into a proton, or will split into two smaller atoms. This process is the basis of any nuclear fission, such as the one that takes place in atomic bombs or nuclear reactors, and it is accompanied by the release of a great deal of

---

[2]This condition is fulfilled mainly when relatively light elements are involved. In very heavy elements like uranium, the number of neutrons exceeds the number of proton.

Box 2.   Carbon-14

Carbon, a chemical element with an atomic number of 6, is one of the elementary components of all organic substances. There are three natural carbon isotopes: carbon-12 (that is, an atom with a nucleus that contains six protons and six neutrons), which constitutes about 99% of all carbon in nature; carbon-13, which constitutes close to 1%; and carbon-14, which constitutes one trillionth of all carbon in nature. While the first two isotopes are stable, carbon-14 is radioactive and decays to nitrogen-14 approximately 5730 years after it is created. Despite this fact, the concentration of carbon-14 in the atmosphere is constant thanks to cosmic rays that collide with nitrogen-14 atoms and cause them to transform to carbon-14 due to the change to a neutron of one of the protons in the nucleus.

Carbon-14 is used to date organic geological and archeological artifacts.

energy and by the emission of nuclear radiation. Unstable isotopes are also called *radioactive isotopes*. Although radioactive materials may be manufactured artificially in a process that requires the investment of energy, such materials also exist on Earth naturally. They are created as a result of the interaction between the cosmic rays coming from space and the atmosphere. A common example is the radioactive isotope carbon-14 (see Box 2). Unstable atoms may also be created artificially in particle accelerators. Such atoms will decay some time after they are created and turn into a different element.

2. Nuclear Density

The average distance between atoms in ordinary matter, like the matter that makes up the Earth and the Sun, is of the order of the atomic radius (which is the distance between the electrons and the nucleus). In other words, the distance between atoms is approximately 100,000 times larger than the size of the atomic nucleus. If the size of an atomic nucleus were about the size of a pinhead, then the average distance between two adjacent atomic nuclei would be about 100 m. This means that ordinary matter is largely porous. When the core of a star collapses and the matter within it compresses, the atomic nuclei and the electrons floating between them gradually crowd together. The average distance between atomic nuclei in white dwarfs decreases to one hundredth that of ordinary matter, while the density increases one million-fold (since the volume decreases as the cube of the distance). If the matter continues to compress, the atomic nuclei will eventually come in contact with each other, yielding a kind of giant nucleus. This is what happens in the cores of neutron stars. However, unlike ordinary atoms in which the number of protons is more or less equal to the number of neutrons, in the case of the neutron star, electrons and protons merge in the core due to the huge pressure that prevails there. Thus, additional neutrons are created, until a composition is obtained that consists mainly of neutrons mixed with a tiny amount of protons and electrons (see Figure 1). The energy required to turn electrons and protons into neutrons is provided by the gravitational potential energy released during the contraction of the core.

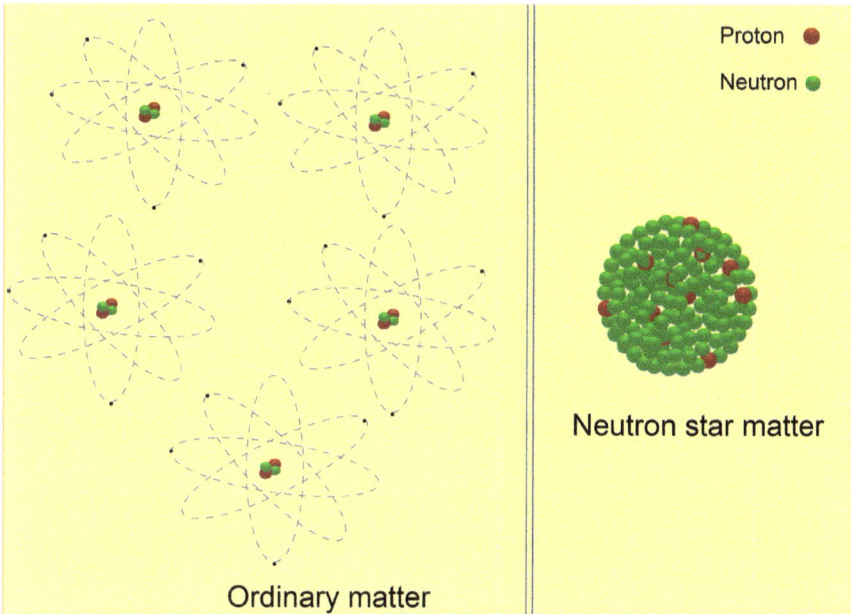

Figure 1. The comparison between ordinary matter and neutron star matter.

It is worth mentioning that although neutron stars resemble giant atoms, there are essential differences between the two: the force unifying the atomic nucleus is the strong force, whereas the force unifying the neutron star is the gravitational force.

How big is a neutron star? Well, the Sun, which is made up of ordinary atomic matter, has a radius of about 1,000,000 km. If we were to shrink the Sun until its density were equal to that of a neutron star, the average distance between its atomic nuclei would decrease by a factor of 100,000. Since the radius of the Sun is proportional to the average interatomic distance, it would diminish by the same factor. Hence, the Sun would shrink and become a sphere with a radius of only about 10 km! Indeed, this is the typical size of a neutron star. Observations, which we will describe later on, have confirmed this.

3. Weight Watchers' Class

The pull of gravity on the face of any star depends on the star's mass and radius. More precisely, the gravitational force is proportional to the star's mass and is inversely proportional to the square of its radius. For example, if we were to double the radius of Earth without changing its mass, the gravitational force on its surface, and therefore also the weight of various bodies on the ground, would decrease fourfold. (The best way to lose weight is, therefore, to inflate Earth....) And if we were to double Earth's mass without changing its radius, the gravitational force would double as well. Since the average mass of a neutron star is about 1,000,000 times the mass of Earth and the average radius is approximately 600 times smaller, the gravitational force on the surface of such a star is about 360 billion times greater than the gravitational force of Earth. If we were to weigh ourselves on neutron stars, we would discover that we have become heavier by that same factor.

This gravitational force imparts great acceleration. A person jumping off a step one meter high would hit the face of the neutron star at a velocity of about 10,000,000 km/h at the moment of impact. It is safe to say that no living creature could survive such an impact. The escape velocity from a neutron star — that is, the minimum velocity required for an object to overcome the gravitational pull of the star[3] — is tens of times greater than the escape velocity from Earth. The escape velocity of a typical neutron star is about one-third the speed of light. Only one kind of object in the universe has higher escape velocities, and these are, of course, the black holes.

4. A Billion Tons on a Teaspoon

Suppose we sent a spaceship to the closest neutron star with the mission of collecting a sample of the matter that makes up that star. What would that sample weigh on Earth? We know that the density of matter of a given mass increases inversely with its volume. Hence, the density of a neutron star is 1,000 trillion[4] times greater than the density of terrestrial matter, and so its weight will be of the same proportion. If we could take a sample of matter from a neutron star, the size of a teaspoon, and preserve its density, it would weigh about five billion tons on Earth. In reality, however, the tremendous pressure of that sample material would cause it to disperse in all directions upon being detached from the neutron star, since there would be nothing to support it. Only on the neutron star itself, where the gravitational force is large enough, can matter exist at such densities.

5. Voyage to the Bottom of a Dense Star

A research delegation journeying from outer space into the bowels of the Earth would encounter a layered structure on its way in. First, the delegation would traverse a very thin envelope of dilute gas called the atmosphere. This layer is only several dozen kilometers thick — about one hundredth of the radius of the entire Earth. After that, the delegation will plow its way through a crust with an average thickness of 30 km. The upper surface of this crust is partially solid (land) and partially water (oceans). Beneath the crust is the upper envelope, which extends from a depth of 30 km to a depth of 400 km. The upper part of this envelope is

---

[3] In technical terms, it is the velocity at which the sum of potential and kinetic energies equals zero.

[4] 1 with 15 zeros, or 10^{15}.

solid and the rest is viscose matter (magma), which sometimes erupts out through the craters of active volcanoes. Continuing on its journey, our delegation will then arrive at a viscose, inner mantle that extends to a depth of 2,900 km. At this depth, they will feel an increase in matter density, mainly due to the large amount of iron present there. Finally, the delegation will reach the core of the Earth, whose radius — approximately 3,000 km — is about half of the radius of the entire Earth, and whose temperature reaches 6,000 degrees Celsius. The upper half of the core is liquid and its center is solid, due to the enormous pressure that exists there.

The layered structure our exploratory delegation encountered in the above description is dictated, to a large extent, by the gravitational force: the deeper we delve, the greater the pressure required in order to support the weight of the upper layers, and the greater the density of the matter, since iron and other heavy elements tend to sink to the center of Earth due to the force of gravity that acts upon them, like the sinking of a stone in water.

A similar voyage to the bottom of a neutron star would reveal a layered structure as well (see Figure 2). Here too, the deeper we go, the greater the

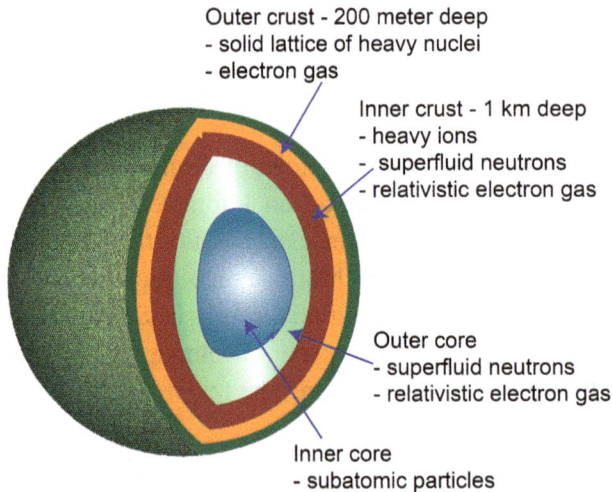

Outer crust - 200 meter deep
- solid lattice of heavy nuclei
- electron gas

Inner crust - 1 km deep
- heavy ions
- superfluid neutrons
- relativistic electron gas

Outer core
- superfluid neutrons
- relativistic electron gas

Inner core
- subatomic particles

Figure 2. The layered structure of a neutron star.

density due to the weight of the upper layers, and here too the structure of the matter changes. However, because of the tremendous gravitational force that exists in the neutron star, the composition of the matter in these layers is completely different from what we are familiar with here on Earth. Seeing as no delegation has yet succeeded in reaching a neutron star, the description of its structure is based on mathematical models, and a great deal of uncertainty exists regarding the detailed properties of the matter found at each depth. The uppermost part contains a very thin layer, about 1m thick, that serves as a kind of atmosphere. The matter in this layer is very thin compared with the matter in the center of the star (but it is still a million times denser than ordinary matter), and its properties are completely governed by the powerful magnetic field that prevails there. Beneath this layer is a solid crust that is about a kilometer and a half thick. The upper part of this crust is made of iron nuclei and perhaps of some lighter elements as well, such as helium and hydrogen, but because of the strong magnetic field there, the shape of the atoms is very different than their shape in terrestrial matter. The inner part of the crust contains extremely neutron-rich nuclei, which under terrestrial conditions are unstable and decay immediately. (Indeed, on Earth, such atoms may be generated only in particle accelerators and only for very short periods of time.) The immense pressure that prevails in the inner crust of the neutron star is what makes these nuclei stable. The pressure is so high beneath the inner crust that it causes neutrons to "leak" from the atomic nuclei (a phenomena called "neutron drip"), and compounds are obtained that contain atomic nuclei and free neutrons with free-roaming electrons in between. As we go deeper, the concentration of free neutrons increases rapidly, and the atomic nuclei become smaller, and eventually disappear. At this radius, the core, whose composition is not entirely clear, begins. The core may be composed of a kind of superfluid that mainly contains free neutrons with very little protons and electrons, and it may also contain different kinds of elementary particles.

6. What is the Maximal Mass of a Neutron Star?

Like in the case of white dwarfs, there is a maximum mass above which a neutron star cannot exist and must collapse into a black hole. The first attempts to calculate this value were made by George Wolkoff and Robert Oppenheimer (who was mentioned in the historical review that opened this book). The value they obtained was smaller than the Chandrasekhar mass, and so they concluded that neutron stars cannot possibly exist in nature. Only later, in 1959, did other scientists show, based on more detailed calculations, that the maximum value is greater than the Chandrasekhar mass and that nothing, in fact, prevents neutron stars from existing in nature. As mentioned earlier, the first neutron star was finally discovered in 1967. Despite the many detailed calculations executed since then, scientists are still debating the exact value of the maximum mass.

The main difficulty involved in accurately calculating this mass lies in the fact that we still lack understanding of the properties of matter at the high densities that prevail in the center of neutron stars. Uncertainty pertains mainly to the degree of pressure that exists within matter subjected to such conditions, and the extent of the pressure opposing the gravitational force is ultimately what determines the value of the maximum mass. Recent estimates, based on advanced models of the strong force, predict that the value of this mass ranges between two to three times the mass of our Sun.

Chapter 12

Quark Stars and Strange Matter

Quark stars and strange stars are still considered to be hypothetical objects that are made of matter referred to as *quark–gluon plasma,* which is, according to conjecture, created in the centers of neutron stars of especially high density. To date, no decisive evidence of the existence of such stars has been brought forward. Recent observations made by the Chandra X-Ray Observatory form the basis of a claim that two objects, which until recently were considered to be neutron stars, are in fact quark stars. The discovery of several especially bright supernovae in the past decade has also led to a suggestion that quark stars, rather than neutron stars, were possibly created in these systems. These claims are controversial and, as of today, no concrete evidence of the existence of quark stars was found, and the search for them is still underway.

1. Quark Soup

Just like an atom is not one single unit but is rather composed of smaller elements — protons, neutrons, and electrons — so are the protons and neutrons themselves not the smallest units of the atomic nucleus; they too are composed of elementary particles called quarks (see Chapter 4). Quarks, like electrons, cannot be further divided and are considered the most elementary units of matter. As mentioned earlier, according to the superstring theory, the particles themselves are a reflection of a more basic unit — the string, but these models, at the moment, are only a conjecture.

The force that acts among quarks is the strong force. It is transmitted through gluons, which are analogous to photons, the modes of the

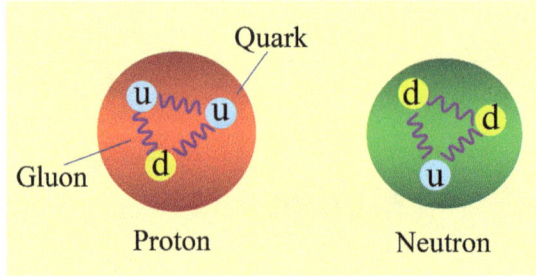

Figure 1. A proton comprises two *u* quarks and one *d* quark. A neutron comprises two *d* quarks and one *u* quark. The forces acting between the quarks are transmitted by gluons.

electromagnetic field. Gluons are the glue that holds the quarks together. There are six known kinds of quarks in nature: the three primary quarks are called *up*, *down*, and *strange* (or *u*, *d*, and *s* for short), and they carry fractional electric charges. Quarks of the types *s* and *d* have a charge equal to one-third the charge of an electron, while *u* quarks have a charge equal to twice the charge of an electron but with the opposite sign. A proton is made up of two *u* quarks and one *d* quark, and so its charge, which is the sum of the charges of its constituent quarks, is equal to the charge of an electron but with an opposite sign. A neutron is made up of two *d* quarks and one *u* quark, and so it is electrically neutral (see Figure 1).

The existence of quarks was predicted both by Israeli physicist Yuval Ne'eman and by American physicist Murry Gell-Mann, who was also awarded a Nobel prize for this discovery. The latter scientist was also the one to give the quark its unique name, one that is reminiscent of a duck's quack and which is borrowed from the Irish author James Joyce's *Finnegans Wake*, in which he wrote: "Three quarks for Muster Mark!". An additional discussion of the name may be found in Gell-Mann's popular book, *The Quark and the Jaguar.* In 1968, several years after quarks were predicted theoretically, they were discovered in experiments conducted at SLAC, the particle accelerator at Stanford University. Since that initial discovery, quarks have been identified many times in various experiments conducted over the years in particle accelerators throughout the world.

The strong force that acts between quarks is like a spring. The more you try to separate two quarks, the greater the attraction between them, similar to a spring — the more you try to extend the spring, the greater the force it applies to oppose that tension. Gluons can be likened to springs that hold the quarks together. Within a proton or a neutron, the quarks are in a relaxed position, like a relaxed spring; but if we try to separate them, this force increases very rapidly. This is the reason it is impossible to find naturally isolated quarks; they are always within protons, neutrons and other subatomic particles that constitute the huge menagerie of hadrons. Protons and neutrons (and hadrons in general) may be likened to a kind of quark prison, in which quarks are serving life sentences.

In the crowded surroundings that prevail in the core of a neutron star, the atomic nuclei merge with one another, and the quarks and gluons from all of the different atomic nuclei mix together, forming a kind of elementary particle soup, called quark–gluon plasma, in which the strong force rules. The properties of such a material are not yet completely known. Scientists are conducting innovative experiments in particle accelerators, like the one shown in Figure 2, in an attempt to create such a compound by causing heavy atomic nuclei to collide. By making comparisons with the theory of the strong force, they are also trying to understand the characteristics of the quark–gluon plasma in much greater detail. As mentioned in Chapter 1, shortly after the universe was created, conditions prevailed that were similar to those found in the cores of quark stars, and the entire universe then contained quark–gluon plasma. Understanding the properties of this substance would, therefore, also advance our understanding of the evolution of the universe and of the conditions that existed in it immediately following the Big Bang.

2. The Strange Matter Hypothesis

Is it possible to create multi-quark compounds that are similar to molecules with a large number of atoms? Researchers believe that multi-quark combinations cannot usually exist. The strange *s* quark is unstable and tends to decay into *d* and *u* quarks. Strange particles, which contain the strange quark and are created in particle accelerators, are therefore unstable and rapidly decay into particles that contain only *d* and *u* quarks.

Nevertheless, some researchers claim that if it were possible to create a multi-quark compound containing an equal number of *d*-, *u*-, and *s*-type

(a)

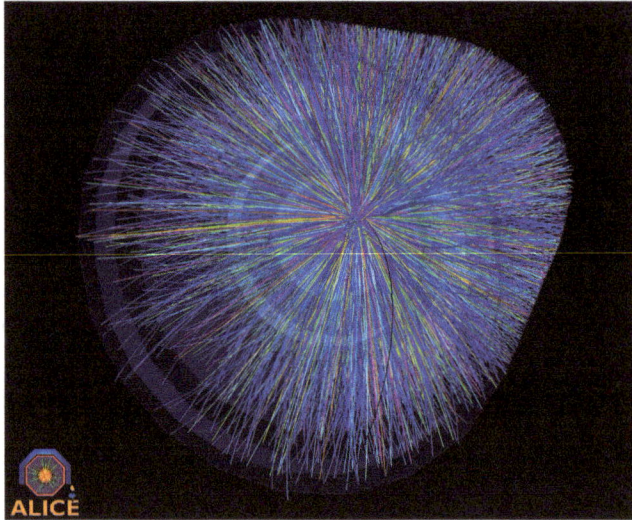

(b)

Figure 2. (b) shows a spray containing thousands of elementary particles created in a collision between atomic nuclei of lead in the LHC particle accelerator. Each line represents the trajectory of a particle measured by the detector. (a) presents a small segment of the accelerator tunnel, containing the pipes in which the particle beams are accelerated. The conditions at the moment of collision that led to the creation of the particle spray were similar to those that prevail in the cores of neutron stars, and in the early universe immediately after the Big Bang. Courtesy of CERN.

quarks, that compound would be stable due to the Pauli exclusion principle. In fact, some even claim that such compounds are more stable than the atomic nucleus, and that all atoms should actually decay and form this kind of strange matter, but because the time that this process takes is so long, it does not occur spontaneously in nature. A deliberate attempt to create pieces of strange matter was made in experiments, conducted in the largest particle accelerators in the world, in which two particle beams that collide at very high energies turn into a sea of quarks for a very short period of time. To date, these experiments have not met with much success. One of the concerns that was raised was that if such attempt were successful, pieces of strange matter created in the accelerator would begin to "digest" ordinary matter and grow, and eventually would destroy the entire Earth. These claims are insubstantial.

If the strange matter hypothesis is correct, then it is expected that strange matter be created in some of the cores of especially dense neutron stars. In professional jargon, such stars are called *strange stars* and their properties are still shrouded by a thick haze. According to one theory, the characteristics of strange stars are very different from those of ordinary neutron stars, and they may be distinguishable in future observations.

Chapter 13

Cosmic Lighthouses

1. Little Green Men

On 28 November 1967, the new radio telescope erected by the astronomy group at The University of Cambridge detected a strange radio signal from a distant region of the galaxy. The signal, which was detected by chance by the telescope, repeated itself with incredible accuracy every 1.337 seconds. It was discovered by a student named Jocelyn Bell, whose doctoral research included a project whose objective was to measure the properties of the interstellar medium using interplanetary scintillations. Had the radio telescope been connected to a speaker instead of to a recording chart, a rhythmic ticking, resembling a clock, would have been heard. Bell's excitement knew no bounds; never before had any periodic radio signal been detected from any astrophysical source. Was she to have the privilege of being the first to detect a transmission from an alien source? Since the source emitting the signal moved over the celestial sphere together with the stars, there was no doubt that the source was extraterrestrial. Bell excitedly shared her discovery with the head of her research group, Professor Anthony Hewish, who was also her doctoral supervisor, and together they playfully decided to call the strange source LGM (which stands for *little green men*), a common nickname for aliens. This important discovery was published on 24 February in the prestigious journal *Nature*.[1] Later on, the object was called a pulsar, which is short for "pulsating source". This name stuck and is used to this day.

---

[1] "Observations of a Rapidly Pulsating Radio Source", *Nature*, 217, pp. 709–713 (1968).

Word of the discovery first reached the local press and was picked up shortly by the international press, igniting the imagination of many. The possibility of radio transmissions from a developed culture located on a planet outside of the solar system was discussed seriously in various circles. Scientists, on the other hand, rejected this possibility after an analysis of the radio signals revealed no evidence of periodic changes in the intervals between the pulses. If the source were located on a planet orbiting around its parent star, like Earth revolves around the Sun, the time between one pulse and the next should have changed slightly as a result of the planet's motion, due to the Doppler effect. Although this change is minute, the detectors available to astronomers at that time were sensitive enough so as to measure such a change if indeed the signal were transmitted from a solar system similar to ours.

In 1974, Anthony Hewish, head of the research group, was awarded a Nobel prize for this discovery, and became the first astronomer (together with Sir Martin Ryle) to win this prestigious prize. The announcement of the prize was accompanied by some discordant notes since Bell, thanks to whom the source was in fact discovered, was not mentioned among the names of the winners.

In the four decades since Bell and Hewish's dramatic discovery, nearly 2,000 additional pulsars were discovered, most of which are located in our galaxy. Their periods range between several thousandths of a second to several seconds. It has become clear over the past two decades that many pulsars emit visible light, X-rays, and gamma rays, in addition to radio emission, a phenomenon that will be elaborated on later.

2. The Lighthouse Effect

What astronomical object can transmit a periodic signal with such unprecedented accuracy? This question fascinated the scientific world immediately after the discovery was published. Less than one year after the pulsar was discovered, an interesting explanation was provided, which later proved to be true. According to this explanation, the source of pulsed emission that Bell detected was a rapidly spinning neutron star that emits two narrow beams of radio waves along a "magnetic" axis that is inclined at some angle relative to the rotation axis. If one of the radio beams sweeps across an observer's line of sight, that observer will see periodic pulsations, like flashes seen from a lighthouse (see Figure 1). Pulsars are therefore cosmic lighthouses. Obviously, as observers located on Earth, we can only see a small proportion of the pulsars in the galaxy — those whose beams sweep across Earth. Most of the pulsars in the universe remain undetected and unknown to us.

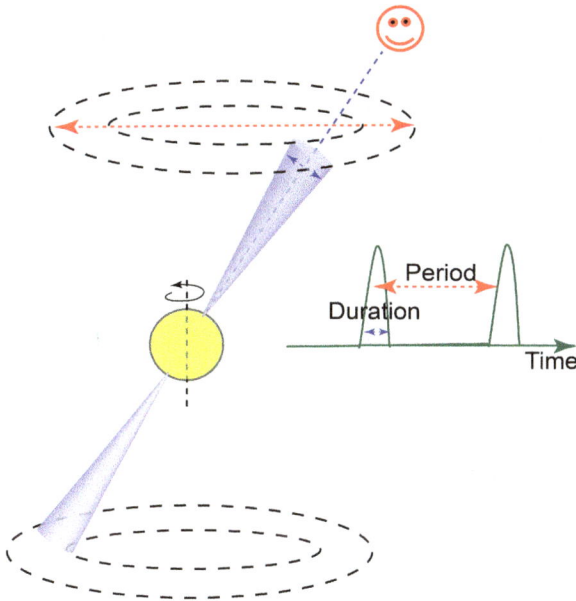

Figure 1. A schematic illustration of a pulsar: The rotation of the star around its own axis creates a lighthouse effect. Observers will see a pulse every time the radio beam crosses their line of sight. The inset on the right presents the shape of the pulses as recorded by the radio telescope.

Box 1. Nature's Most Precise Clocks

Pulsars may be used as clocks: The rotation period of pulsars may be used as a time basis. The extraordinary accuracy of the periodic pulsations makes pulsars the most accurate clocks in the universe, more accurate even than the most advanced atomic clocks.

How can we be sure that a pulsar is indeed a neutron star, and not a white dwarf or just an ordinary star? First, if it were an ordinary star, it could be seen in visible light using an ordinary telescope. In most cases, there is no evidence of any stars in the direction of the pulsar. Second, if an ordinary star were to rotate too fast, its external parts would tear off by the centrifugal force that acts on them, and the star would be destroyed, somewhat similar to mud that is thrown off a vehicle's tires, flying in every direction, as the vehicle accelerates. Our Sun, for example, can rotate around its axis at a maximum frequency of one revolution every three hours before it starts coming apart. In reality, the Sun completes one revolution every 30 days more or less. For a white dwarf, the maximum rotation frequency is one revolution every 10 sec (or 0.1 Hz). The short periods of pulsars — several milliseconds[2] in extreme cases — requires the star to have a much smaller radius than that of a white dwarf. The only thing we are familiar with that fits that requirement is a neutron star. As we described extensively in the first part of this book, neutron stars are the densest objects in the universe that are not black holes. A typical neutron star has a radius of "only" about 10 km, and it has a tremendous gravitational force that can offset the centrifugal force that acts on its surface even when the star is revolving at much higher frequencies than would cause any other object to tear apart. To be precise, a typical neutron star can approach a rotational period of a few milliseconds before its outer part starts flying apart — frequencies that are higher than that of any pulsar discovered to date.

3. Magnetic Tops

Two components are required in order to produce the radio waves that pulsars emit — rapid rotation and a strong magnetic field. We already presented observations that indicate that the rotation frequency of pulsars can exceed several revolutions per second and, in extreme cases, may even come close to 1,000 rev/sec. Measurements, which we will describe shortly, lead to the conclusion that the magnetic field on the face of a pulsar is 100 billion times or more stronger than that of an ordinary household magnet. What, then, is the origin of the neutron stars' rapid rotation and huge magnetic fields?

---

[2]A rotational period of 1 msec is equivalent to 1,000 rev/sec.

Box 2.   The Coriolis Force

An apparent force is a force that we seemingly measure when we are in an accelerating system. For instance, when we accelerate the car we are driving, we feel a kind of force that "pushes" us backward. This force is an apparent force and it does not stem from any physical force that is acting on us, but rather from the fact that the system we are in, in this case our car, is accelerating.

Such a force is felt on Earth, as a result of Earth's rotation, and is called the Coriolis force, after French scientist, Gaspar Coriolis, who was first to describe it. The Coriolis force causes cannonballs to deviate from their trajectory and cyclones to swirl, as well as many other phenomena, some of which are familiar to us all. Fire control systems in long-range cannons and flight route computers on jet planes take this effect into consideration.

The answer to this question relates to the properties of ordinary stars. Most stars in the universe revolve around their axis, including the Sun, as mentioned earlier. Most of the stars also have a magnetic field. According to the dynamo theory, the magnetic fields of stars and planets, including Earth, are created by electric currents that are induced within the star as a result of the differential rotation of different layers of the star. In the case of Earth, the electric currents are formed in the outer part of the Earth's core, which contains liquid iron and nickel. Due to an apparent force called the *Coriolis force* (see Box 2), the liquid matter that surrounds the core moves at a slightly different angular velocity than the inner part of the core, and since the rotating liquid has a high electrical conductivity, electric current is induced in it. A similar mechanism is present in the Sun and in other stars, where non-uniform rotation of electrically conductive ionized gas creates electric currents that move in a circle around the center of the star and induce a magnetic field similar to that of a household magnet. Special photographs of the Sun, like the one in Figure 2, show the magnetic field lines leaving the face of the Sun, illuminated by hot gas that flows along them.

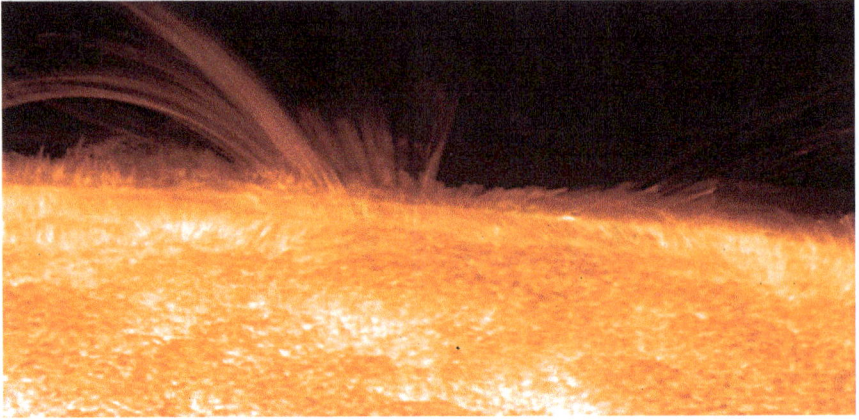

Figure 2. The face of the Sun as photographed by the solar optical telescope aboard the Hinode satellite. The reddish lines leaving the face of the Sun are the magnetic field lines, which are illuminated by hot gas that is flowing along them. Image Courtesy Hinode JAXA/NASA.

After a heavy star consumes all of its nuclear fuel and there no longer exists a heat source that can supply the pressure required to offset the gravitational force, the inner part of the star begins to collapse, while its outer parts are ejected in all directions, creating a supernova. During the collapse process, the core of the collapsing star gradually shrinks, becoming denser and denser, until it is so dense that the quantum pressure of the neutrons prevents any additional collapse, and a neutron star is created. The core, like the entire star, is magnetized and revolves around its axis; the smaller the core, the greater its angular velocity, as stems from the law of conservation of angular momentum. This is similar to the way the rotational speed of a spinning ballet dancer increases when he draws his outstretched arms to the sides of his body while spinning. Applying the law of conservation of angular momentum, it can be shown that increase in angular velocity is inversely proportional to the square of the radius of the collapsing core. Since the radius of the neutron star is about 100,000 times smaller than the radius of the star that created it, its angular velocity is about 10 billion times greater than that of the original star. For example, if the Sun were to collapse to the size of a neutron star, its rotation rate would increase from one revolution per 30 days to about 1,000 rev/sec. This explains the tremendous rotational speed of pulsars.

When the core of a star collapses, the star's magnetic field also compresses. The high electrical conductivity of the core prevents the magnetic field lines from escaping, and since the intensity of the magnetic field is proportional to the number of field lines per unit area, the magnetic field strength increases inversely with the surface area of the star. Hence, the magnetic field strength, like the angular velocity, is 10 billion times higher than that of the original star. After a typical star, like our Sun, collapses, the values obtained for its magnetic field correspond to those measured for pulsars (see Figure 3).

How do the rapid rotation and the tremendous magnetic field affect the neutron star? The surface of the neutron star contains charged particles — positively charged ions and electrons. The powerful magnetic force that acts upon these particles prevents them from crossing magnetic field lines

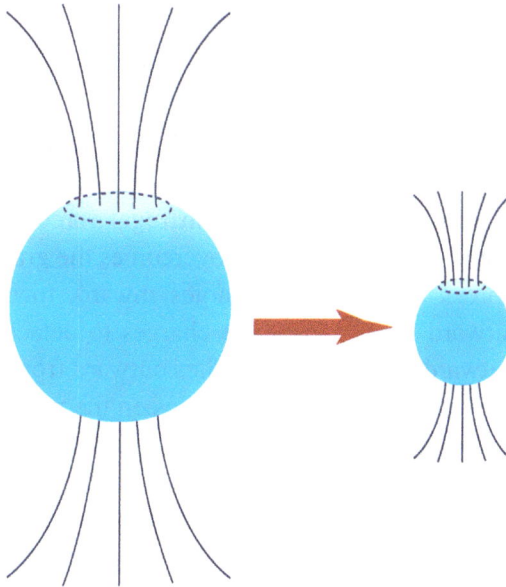

Figure 3. Collapse of a magnetized star: The strength of the magnetic field in the area denoted by the dashed-line circle is proportional to the density of the field lines within the circle, or more precisely, to the number of lines per unit area. Since the number of field lines crossing the surface of the collapsing star must remain constant, when the star shrinks, as the illustration demonstrates, the lines become more crowded, and the magnetic field strength on the stellar surface increases.

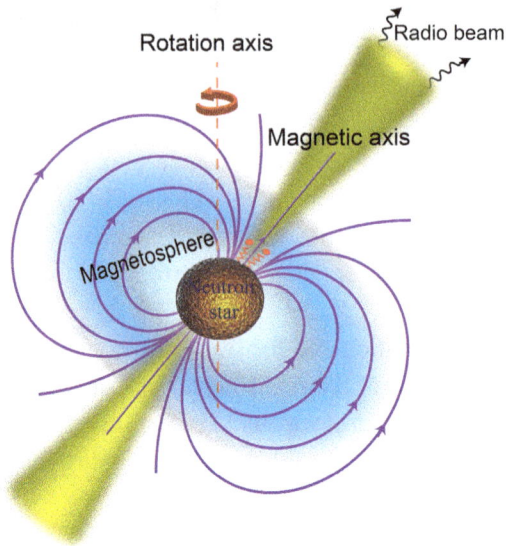

Figure 4. The structure of a pulsar.

and enables them to move only along these lines, similar to the motion of a bead on a wire. When the magnetic axis is misaligned with respect to the rotation axis (see Figure 4), the centrifugal force acting upon the charges due to the rapid stellar rotation overcomes the gravitational force and the other forces that pull the charges inward, toward the center. The resultant outward force causes the charges to detach from the star and accelerate outward, like a bead that flies outward from a rapidly spinning wire. The free charges accumulate between the magnetic field lines above the stellar surface and form a layer of very thin, magnetized gas comprising ions and electrons, called the *magnetosphere*. The magnetosphere corotates with the entire star, and the velocity of its outer parts approaches the speed of light. Since nothing can move faster than the speed of light, the inertia of the trapped matter causes the magnetic field lines leaving a small region close to the magnetic poles to twist and stretch. These lines, which would be closed in the absence of stellar rotation, tear open to infinity, and charges are thus free to move to regions that are very distant from the star. The charges that are accelerated along the open field lines form narrow particle beams that travel outward from the edge of the star at a velocity that approaches the speed of light. These charged particle jets are responsible for the beamed radio waves that pulsars emit. In addition, because the magnetic axis and the rotation axis are misaligned, magnetized matter is also ejected along the equatorial plane in the shape of a torus (a doughnut shaped cylinder) that envelops the pulsar. This matter too travels at a speed that approaches the speed of light and is called a *pulsar wind*.

Where does the energy emitted by the pulsar come from? This energy is, of course, rotational energy. The radiation the pulsar emits slows it down, just like the friction between a wheel and its axle causes it to decelerate as rotational energy is converted to heat at the points of friction. The pulsar's rotation frequency gradually decreases over time, resulting in a change in the pulsar's cycle time, as seen by the observer. This change is very slow, but thanks to advanced radio detector technology it can be measured. Measuring the pulsar's rotation period and the change in the rotation period enables scientists to infer the pulsar's age, to calculate its magnetic field strength, and to measure the net rotational energy that the pulsar loses over time. The magnetic fields strengths measured using this method are indeed compatible with the values expected when a star collapses. Measuring the rotational energy reveals that the energy emitted in the radio band is only a small fraction, about one millionth, of the rotational energy extracted from the star. Most of the rotational energy is used to accelerate the pulsar wind and is converted into the kinetic energy of that wind. A substantial part is also emitted in the form of gamma rays.

The pulsars discovered so far range in age between several hundred years and several tens of millions of years. This datum is slightly surprising, considering the fact that the galaxy in which we live is about 100 times older. Why have older pulsars not been discovered, as one might expect? The reason is simply that when the rotation speed drops below a certain critical value, the centrifugal force decreases to a level that does not enable additional charges to detach from the surface of the star, the ejection of charged particle beams ceases, and the pulsar is extinguished. Even though magnetized neutron stars continue to rotate, they can no longer emit radiation, and they disappear into the darkness of the universe, like lighthouse lamps whose power supply has been cut off. In fact, most

Box 3.   Powerful Masers

The properties of the radio waves a pulsar emits teach us that the mechanism producing this radiation is in fact a giant radio laser called a maser (a maser is a laser that emits radiation in the radio and microwave ranges). A pulsar's radio beam is in fact the largest maser in the universe. Additional masers exist in astrophysical systems that contain molecular gas, but these have a different *modus operandi*. The way a pulsar maser works is not yet entirely understood. One possibility that has been proposed is that the charges flowing along the open magnetic field lines are bunched on scales that are smaller than the wavelength of the radio waves. Each charge bunch emits coherently like a single gigantic charge. This method serves us on Earth in a device called a free-electron laser. Other theories also exist, and it looks like some time will go by before the pulsar's radio emission mechanism is understood properly.

neutron stars in the universe are no longer active; indeed, the universe is a giant graveyard for pulsars that have died of old age.

Advanced pulsar models developed over the years predict a connection between the strength of the neutron star's magnetic field and the pulsar's lifetime (or alternatively, the critical spin period required to activate a pulsar). Measurements conducted on hundreds of pulsars have confirmed that such a connection (called a "pulsar death line") indeed exists, providing important verification of the pulsar theory.

4. The Crab Nebula and the Crab Pulsar

One of the most famous pulsars resides in the center of a gaseous nebula within the Taurus Constellation, called the *Crab Nebula*. This nebula is located at a distance of 6,000 light years from Earth, its diameter is 11 light years, and its outer part is expanding at a velocity of 1,500 km/sec. Matter in the nebula's interior is seen to be traveling at a velocity that is close to half the speed of light. The Crab Nebula was created in a supernova explosion that occurred a long time ago, and it constitutes a kind of memorial for the star that lost its life in that explosion (see Figure 5).

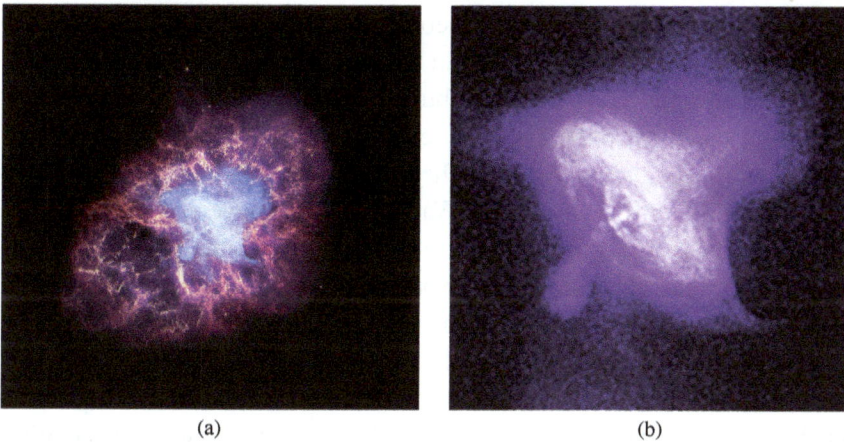

(a)   (b)

Figure 5. (a) Integrated photo of the Crab Nebula. The blue region in the center is an X-ray image. The yellow and red are more outer regions that were photographed in visible light, and the purple is an infrared image. The different photos are superimposed so as to demonstrate the structure of the nebula. (b) Magnified X-ray image of the inner region. The pulsar is located at the center of the white spot from which the jet is emitted. The rings that are visible in the X-ray image are emitted from a magnetic torus that envelops the pulsar, which constitutes the pulsar wind. The diameter of the inner ring depicted in this photo is one light year. X-ray: NASA/CXC/SAO/F. Seward; Optical: NASA/ESA/ASU/J. Hester & A. Loll; Infrared: NASA/JPL-Caltech/Univ. Minn./R. Gehrz.

The explosion that led to the birth of the Crab Nebula was observed by Chinese, Arab, and North American Indian astronomers in July 1054. According to Chinese reports, the supernova could be seen with the naked eye during the daytime for a period of three weeks, and during the night for close to two years. Its brightness approached that of the moon. One can imagine the surprise of the Chinese astronomers when one night, they suddenly discovered an especially radiant spot of light at a point in the sky where no star was previously seen. Indian drawings, discovered in cave in North America, depicted a huge star with rays of light radiating from it, evidence of the dramatic event.

The Crab Nebula was first discovered in 1731 by the British astronomer John Bevis. The nebula was observed again in 1758 by Charles Messier while he was searching for comets, and it is classified as the first item in his catalog. The nebula was named only in the 1840s by Irish astronomer William Parsons, 3rd Earl of Rosse, after observing it using a telescope he had built.

The connection between the nebula and the supernova observed by the Chinese in 1054 was proposed in the early 1900s, when an analysis of early photographs showed the nebula to be expanding. Based on the expansion rate, astronomers then calculated the point in time when the process began, which matched the time of the explosion as reported by the Chinese. Later on, astronomer Walter Baade discovered that a tiny star indeed exists in the center of the nebula, at the same location where, according to the calculations, the expansion began. It later turned out that this star is the pulsar that powers the nebula.

The Crab Nebula is among the most observed objects in modern astronomy. Observations from countless telescopes and satellites conducted over the years show that the nebula is a strong source of radiation of an extremely broad spectrum — from radio waves to gamma rays of the highest energies. Similar radiation was discovered also in the Vela Nebula and other similar nebulae, all of which have a pulsar in the center of the nebula. What mechanism in these nebulae is responsible for the emission of this radiation? Detailed models of the nebula structure establish the assumption that the radiation is emitted from a strong shock wave created in the collision between the pulsar wind and the dense matter around the pulsar. This matter made up the envelope of the star that exploded and created the nebula, and it is denser than the average density of the interstellar medium located throughout the entire galaxy. The thin, fast pulsar wind, which is thrust from the neutron star, moves outward, until it collides with the dense matter that surrounds it. This may be likened to a race car crashing into a concrete wall or sea waves breaking against a breakwater, with the difference that the pulsar wind is emitted in a continuous manner. The powerful collision creates a shock wave in the central region, near the innermost ring visible in the above X-ray image, like a shock wave created when air collides with an airplane flying at supersonic velocities. The matter that passes through the shock wave is heated to very high temperatures that lead to the observed nebular emission.

5. James Bond and the World's Largest Pulsar Observatory

In *GoldenEye*, the 17th movie in the James Bond series, Bond attempts to locate a satellite dish that activates a special satelite named GoldenEye, which is capable of generating a powerful electromagnetic pulse that can destroy any electronic device within a radius of many kilometers from the point of impact. After many hardships, Bond and his counterpart, Simonova, manage to penetrate the facility's operation center, disrupt the activation of the satellite, and destroy it along with the dish. In one of the movie's final scenes, Bond struggles with Agent 006, who has betrayed his country and has joined forces with the Russian Mafia. The battle between the two takes place on the huge satellite dish that was used to communicate with GoldenEye; Bond ultimately overpowers his adversary. This scene, like other scenes in the movie, was shot in Puerto Rico, near the town of Arecibo. The dish featured in the movie was not built specially for the film but was already there (see Figure 6). Its main role was, and still is, to act as the world's greatest pulsar observatory. The movie was

Figure 6. The radio dish at the Arecibo Observatory located near Arecibo, Puerto Rico. Courtesy of the NAIC — Arecibo Observatory — Puerto Rico, an NSF facility.

filmed during a time when renovations were being conducted at the facility with the objective of improving the reception quality of the radio dish. The radio dish at the Arecibo Observatory was featured in other movies as well, the most famous of which is *Contact* starring Jodie Foster.

The diameter of the Arecibo dish is 305 m and it is, in fact, used as a radio wave mirror. Radio waves are reflected off the mirror to receivers located on a 90-ton platform that is suspended 150 m above the mirror. The immense area of the radio dish, approximately 20 acres, enables it to receive very faint signals and makes it the most sensitive radio antenna ever constructed. Aside from detecting pulsars, the Arecibo Observatory is used for various astronomical radio observations, to study the atmosphere (aeronomy), to view bodies in the solar system using the radar astronomy method, and for the SETI project (see Box 4).

Many important discoveries were made over the years at the Arecibo Observatory, including the discovery of a system of binary neutron stars that emits gravitational waves — a discovery that will be discussed shortly.

6. New Discoveries — Gamma Radiation from the Magnetosphere

The launch of the Fermi Gamma Ray Satellite in 2008 led to the discovery of a new family of pulsars whose main characteristic is the emission of strong gamma ray pulses. The interesting thing about these pulsars is that many of them do not seem to emit any radio waves, while for those that do emit radio waves, no clear connection can be found between the gamma ray pulses and the radio pulses. One explanation offered is that the gamma ray and radio beams are emitted from different regions of the magnetosphere and in different directions, so that in pulsars in which no radio pulses are seen, the radio beams are directed so that they do not cross our line of sight. In pulsars in which both radio and gamma ray pulses are seen, the two beams cross our line of sight, but not at the same time, so there is no clear connection between the two signals. The analysis of the gamma ray characteristics in the sample of pulsars discovered so far by Fermi provides invaluable information regarding the structure of the magnetosphere and enables astronomers to map the different processes that take place there.

Box 4.   The SETI Project

The Search for Extra Terrestrial Intelligence (SETI) Project is the name given to the attempts to find intelligent life outside of Earth. Pioneering these attempts was astronomer Frank Drake, who in the early 1960s, following an article he read in *Nature*, began looking for evidence of intelligent radio signals from outer space using the radio telescope at the Green Bank Observatory in West Virginia. Since then, many attempts have been made, but so far none have met with any success.

In 1974, Drake and astronomer Carl Sagan (who gained fame also as the author of popular science books), initiated the transmission of encoded radio message toward the globular cluster M13, 25,000 light years from Earth. The message was transmitted using the radio telescope at the Arecibo Observatory and it included various information, such as the numbers 1–10, the structure of DNA, a graphic illustration of the solar system, and so on. Since the message will travel some 25,000 years from the day of its transmission until it reaches its destination, this attempt may be regarded as a demonstration of the human technological abilities rather than as a serious attempt to contact aliens.

Over the years, NASA has funded a variety of studies related to the SETI Project. In 1992, NASA decided to allocate funding for a long-term project that was to include systematic sky surveys and imaging of hundreds of nearby stars using a network of radio telescopes. The project was canceled one year later due to congressional objection to funding projects of this kind. Since then, the SETI Project has been funded mainly by private entities. The movie *Contact* is based on a Carl Sagan book that was inspired by this project.

7. Neutron Star Tango and the Theory of General Relativity

In 1974, Joseph Taylor and Russel Hulse, two astronomers from Princeton University,[3] discovered a new pulsar while conducting observations at the Arecibo Observatory. The pulse period measured in the initial analysis of

---

[3] At that time, Taylor served as professor of physics at the University of Massachusetts and as Hulse's doctoral supervisor. Only later did the two relocated to Princeton University.

the data indicated a typical neutron star with a rotation frequency of 17 cycles per second. After more extensive observations, however, Taylor and Hulse noticed a strange phenomenon the likes of which had never been seen before. The pulsation period, in other words, the time interval between two consecutive pulses, was not fixed like for all other pulsars discovered until then, but rather it varied systematically in a cyclic manner — sometimes the pulses arrived earlier than expected and other times later. After meticulous analysis of the data, it became clear that the modulation of the pulse period — in other words, the change in the arrival times of the pulses — occurs in a cycle of 7.5 hours. The most likely explanation for this phenomenon is that the pulsar itself revolves around another star in its vicinity, and that the cyclic change in pulse arrival times stems from a Doppler effect due to the pulsar's motion relative to the observer: When the pulsar moves away from us, the time interval between the pulses grows, and when it moves toward us, it shortens. The short orbital period of the binary pulsar, about 7.5 hours (for the sake of comparison, it takes Earth one year to complete a revolution around the Sun), indicates that the distance between the pulsar and its companion star is very small. Since the search for an ordinary star in the pulsar's direction produced no results, the question arose: Who is the pulsar's companion? After analyzing the pulsar's orbit, Hulse and Taylor finally concluded that the companion must also be a neutron star, and that in fact the system was a binary system of neutron stars of similar masses.

How was this system created? Unlike the Sun, most of the stars in the universe exist in pairs, that is, as part of binary systems that include two ordinary stars in orbit around each other. If the stars' mass is big enough, they will become, each in its turn, black holes or neutron stars. An obvious question might be, what will happen to the binary system when the time comes for the heavier partner, whose life span is shorter, to die? Well, in some of the cases, the explosion of the star will cause the system to break up and the two stars will separate. The exploding star will become a dense object, while the other star will continue to roam around freely throughout the universe until its foreseen death. In other cases, however, in which the binary system is stable enough, the stars will remain intact, and a new binary system will be formed in which an ordinary star and a reincarnated star, in other words a black hole or a neutron star, exist side by side. At some point, the time will come for the lighter partner to end its life as well, and in most cases, it will become a white dwarf, resulting in a binary system in which a white dwarf revolves around a neutron star or a black hole. If, however, the companion's mass exceeds 10 solar masses, it will explode and leave behind a neutron star or a black hole. The binary system will then consist of two black holes, or a black hole and a neutron star, or two neutron stars revolving around each other in a kind of celestial tango. In the system that Hules and Taylor discovered, after ending their lives, the two ordinary stars had become neutron stars. According to different estimates, about 5% of all neutron stars in the universe are in binary systems, and the rest are single.

Immediately upon its creation, the neutron star in a binary system becomes a pulsar and starts transmitting pulses. If the time that has passed since the creation of the binary neutron star system is shorter than the pulsar's lifetime (some 100 million years), at least one of the stars will still be emitting radio pulses. Observers who are lucky enough to be in the path of one of the radio beams, will then discern a pulsar with a pulse period that changes periodically according to the pulsar's orbital frequency (Figure 7). This is exactly what Hulse and Taylor discovered. On the other hand, if the time that has passed from the creation of the binary system exceeds the pulsar's lifetime, the pulsars turn off, and what will remain will be a binary system of dead pulsars (or neutron stars that do not pulsate). Researchers

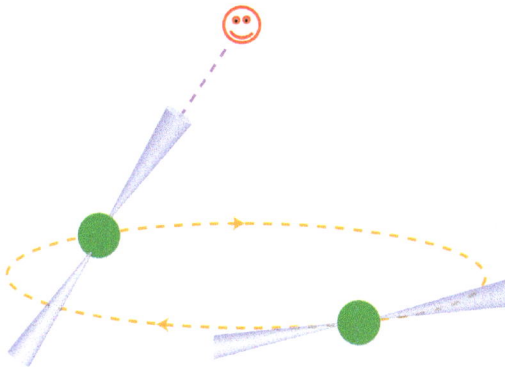

Figure 7. A binary neutron star system: One of the two stars, or both, as the illustration demonstrates, may be a live (active) pulsar. The radio beams from the two pulsars lie usually on different planes, so that a distant observer will see only one pulsar.

estimate that the universe contains a large number of such objects. Since they are inactive, they cannot be detected using existing means. However, as we will explain shortly, we may be able to discover some of them in the future thanks to the gravitational waves they emit.

The discovery of the binary neutron star system provoked reactions in the scientific community, but the truly great importance of the discovery became clear only several years later. Hulse and Taylor continued to devoutly monitor the pulsar they discovered, and every couple of months they conducted additional measurements and reevaluated their data. To their surprise, they found that the distance between the two neutron stars is not constant as they expected, but rather decreases over time. The change in this distance was so small that it took them several years to notice it. Detailed calculations made following this finding showed that the change is the result of the emission of gravitational waves due to the orbital motion of the stars, as Einstein's theory of general relativity predicts. As we explained in previous chapters, the presence of the neutron stars causes the space around them to warp. If the stars were stationary, the space warping would have been constant in time. Because of the stars' motion around each other (and more precisely, around the common center of mass), the warping pattern changes over time in a periodic manner, and the ripples of the spacetime propagate through the universe, like waves that are created when two iron spheres are rotated around each other in a large water bath (see Figure 8).

The agreement between the theoretical prediction and the measured rate of change of the orbital separation between the stars was almost perfect, as Figure 9 depicts. The pulsar was monitored for over 30 years, and the measurements confirmed the theory unequivocally. This was the first time that gravitational waves were observed, even if indirectly, providing additional, and most importantly confirmation of the correctness of the theory of general relativity.

This discovery earned Taylor and Hulse a Nobel Prize in 1993. This time, the committee was wiser and avoided the mistake it made in the case of Bell and Hewish; Hulse, who was a doctoral student and worked under the guidance of Taylor at the time of the discovery, was included in the list of prize winners. As mentioned in Part 2 of this book, the first direct detection of gravitational waves from a binary black hole system was made by LIGO in September 2015. The first direct detection of gravitational waves from a binary neutron star system was made on August 2017, more than

Figure 8. A pair of rotating neutron stars emitting gravitational waves. Courtesy of INFN/made by Luca Cesaro.

four decades after the discovery of the Hulse–Taylor pulsar. The development of LIGO was inspired in part by observations of the Hulse–Taylor pulsar.

Several additional such systems have since been discovered. The high precision that is achievable when measuring pulsar emission enables scientists to examine a few other effects predicted by the theory of general relativity, and so pulsars in binary systems serve as laboratories for testing the theory of general relativity. In 2004, a system was discovered in which the beams of both pulsars were detected (double pulsar). Since both beams are in the same plane, an eclipse occurs when one of the pulsars is exactly behind the other (like in a solar eclipse). By measuring the eclipse of the radio emission it was possible, for the first time since the discovery of the first pulsar in 1967, to directly map the structure of the pulsar's magnetosphere and to learn a great deal about its emission mechanisms. Many theoretical efforts in this direction are currently underway.

What will be the ultimate fate of the neutron star system that Hulse and Taylor discovered? Well, the intensity of the gravitational radiation emitted by a binary neutron star system increases with the decrease in the

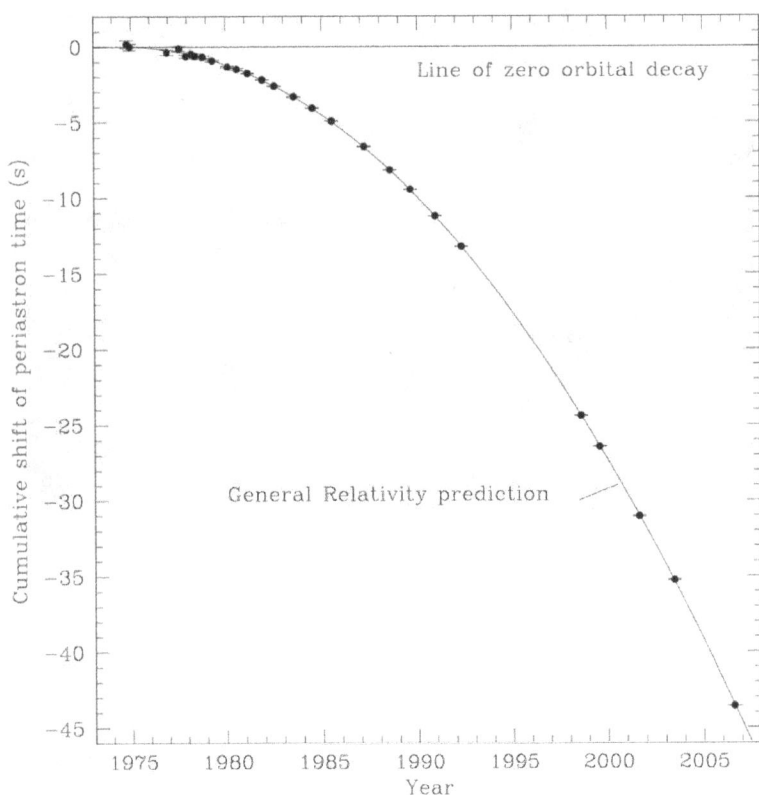

Figure 9. The original measurement of the orbital decay of PSR B1916+16, the pulsar discovered by Hulse and Taylor. The horizontal axis represents the time (in years) since the pulsar was discovered. The vertical axis reflects the change in distance between the two neutron stars. A negative value means that the orbital separation is smaller than it was in 1975, when the pulsar was discovered. Each dot indicates a measurement. The line on which the dots are plotted is the theoretical prediction of the theory of general relativity. Reproduced with permission from Weisberg, J.M., *et al.*, *ApJ*, 722, p. 1030, (2010).

orbital separation. The emission of gravitational waves causes the distance between the stars to shorten over time, as Hulse and Taylor's discovery demonstrates, and as a result, the emitted gravitational waves intensify. This process ends when the two stars collide and crush in an explosion that appears as a gamma ray flash, as will be elaborated on later. This is how Hulse and Taylor's PSR B1916+16 system will, supposedly, end its life in approximately 300 million years. We most probably will not live to see the collision of the neutron stars in this system, but similar systems exist in nearby galaxies in the universe in which the neutron stars are about to collide in imminent years. Shortly before the collision, the intensity of the emitted gravitational waves is expected to be so high that, scientists estimate, it will be directly detectable using the LIGO detector, described in detail in Chapter 8. Indeed, a burst of such gravitational radiation, accompanied by a gamma ray flash, was finally discovered in 2017, and scientists hope to detect additional, similar events in the near future. Each of these events serves as a record of the final stage of a binary neutron star system somewhere in the universe. The data encrypted in the gravitational waves will enable scientists to estimate the neutron stars' mass and to learn more about their nature.

8. X-Ray Pulsars

In binary systems in which a neutron star is a companion of an ordinary star, matter overflows from the ordinary star to the neutron star due to the strong gravitational force it induces in its surroundings. If the neutron star's magnetic field is strong enough, the accreted matter will flow through the magnetic poles, since it cannot cross the magnetic field lines, but flow along them. The flowing matter that is compressed at the edge of the neutron star, in the polar region, becomes hotter, and emits X-rays. In fact, X-rays are emitted as beams from two "hot spots" near the star's poles, like the radio beam emitted by an ordinary pulsar. The neutron star's self-rotation causes the lighthouse effect, and the X-ray emission reaches the observer in the form of pulses. These objects are therefore referred to as *X-ray pulsars*. X-ray pulsars do not emit radio waves, since the large quantities of matter that flow through the magnetic "funnel" near the poles quench the radio emission mechanism. It is worth noting that in some of the X-ray pulsars discovered, the angular velocity of the pulsar increased over time, contrary to the tendency exhibited by ordinary pulsars. The increase in angular velocity apparently results from the accretion of the matter by the pulsar.

9. Cosmic Starquakes

As we mentioned earlier, pulsars are very accurate clocks. With the exception of a very slow waning in the rotation frequency, which stems from the loss of rotational energy, the time interval between one pulse and the next is fixed. Nevertheless, a sudden change is occasionally seen in the pulse periodicity, which can be interpreted as a sudden increase in the pulsar's rotational frequency, after which the system gradually relaxes. This phenomenon, called *a glitch*, is relatively rare. It is not seen in all pulsars, and those that do exhibit it, do so only seldomly. In most cases, the change in the star's rotational frequency during the glitch is very small, about one to a million, but it occurs within a very short time period. The glitch is usually followed by a recovery phase that lasts from days to months, during which the rotational frequency gradually returns to the pre-glitch value. Although the change in the periodicity is small, it requires the release of a significant amount of energy within the short time period during which it occurs. What causes this phenomenon?

According to the accepted conjecture, such a change in a pulsar's rotational speed may be caused when the solid crust of the neutron star is slightly deformed and undergoes rearrangement, similar to what happens during an earthquake on Earth, when the displacement of Earth's tectonic plates sometimes causes its surface to vibrate. Like Earth's solid tectonic plates that float around on a layer of viscous liquid, the solid crust of the neutron star rests upon the core, which contains a neutron superfluid and rotates at a slightly higher velocity than does the solid crust. For a reason that is still not completely understood, the liquid core and the crust are suddenly coupled, and as a result, the crust is deformed and its rotation abruptly accelerates. After this event passes, the crust resumes rotating at its original velocity. The coupling is probably caused by a sudden change in the magnetic field that penetrates both the core and the crust.

Chapter 14

Magnetars — The Strongest Cosmic Magnets

On 27 December 2004, several research satellites launched by NASA and other space agencies detected a huge burst of gamma radiation. The event, which lasted less than one second, was the result of a sudden *magnetic field reconfiguration* in a highly magnetized neutron star located in a distant region of our galaxy, about 50,000 light years from Earth. It was the strongest explosion documented in the Milky Way galaxy in the past 1000 years (much stronger explosions occur throughout the entire universe, as will be explained later on). Since most of the explosion energy was emitted in the form of gamma rays, it was not visible to the naked eye, but the luminosity released during the explosion was so great that it "blinded" the gamma ray detectors installed on the various satellites and caused deviant oscillations in the Earth's atmosphere. Had this explosion taken place closer to Earth, its effect would have been similar to that of a large nuclear bomb, and it might have possibly annihilated all life on Earth.

The star that caused the 2004 burst, which is called SGR 1806-20, belongs to a class of objects called *magnetars*. Magnetars are neutron stars with magnetic fields that are much stronger than those of the ordinary pulsars described above. Indeed, they may be 1,000 times, and more, stronger than the magnetic field of an average pulsar. SGR 1806-20 is the object with the strongest magnetic field measured thus far in the universe. The burst in 2004 was not, however, the first of its kind. Several satellites had measured a similar eruption from a source in a galaxy adjacent to the

Milky Way, called the *Large Magellanic Cloud*, as early as 1979, and later on, several smaller eruptions with similar characteristics were measured as well, but their nature was unclear. At first, these eruptions were thought to be gamma ray bursts of the kind discovered in the late 1960s by the Vela satellites (see Chapter 18), but it was subsequently discovered that these eruptions have different characteristics, and that suggestion was rejected. To distinguish this population of bursts from regular gamma ray bursts, they were called *soft gamma repeaters*. Only in 1992 was a theoretical explanation of this phenomenon offered, and a few more years passed before the existence of magnetars was confirmed observationally. There are now over a dozen known magnetars. See Table 1 for a list of sources of magnetic fields.

At the same time as the discovery of soft repeaters, another family of objects was discovered, which were named *anomalous X-ray pulsars*. These are pulsars that emit X-rays instead of radio waves (and are, therefore, anomalous). Like the ordinary pulsars described in Chapter 13, the X-ray emission comes in pulses that attest to a rotating neutron star. Measurements show, however, that the magnetic field of these stars is 100

Table 1.   Sources of magnetic fields.

Intensity (Gauss)	Source	Notes
0.6	Earth's magnetic field	Measured at the North Pole
100	Household magnet	Like a refrigerator magnet
1,000	The magnetic field of strong Sun spots	The average field on the face of the Sun is about 1,000 times weaker
4.5×10^5	The strongest magnetic field ever created in a lab	Achieved using superconductors
10^8	The magnetic field measured in white dwarfs	
10^{12}–10^{13}	The range of magnetic field strengths measured in ordinary pulsars	
10^{14}–10^{16}	The range of magnetic field strengths measured in magnetars	

times, and more, stronger than that measured for ordinary pulsars. For many years, researchers failed to see the connection between the anomalous pulsars and the soft gamma repeaters. Only after advances were made in the theoretical understanding, following the work of physicists Christopher Thompson and Robert Duncan in 1992, was it proposed that the two populations originate from the same physical object: the magnetar. This suggestion gained wide support several years ago, after a gamma ray burst with characteristics similar to those of soft gamma repeaters was measured in one of the anomalous X-ray pulsars.

The structure and mechanism of magnetars have not yet been fully revealed. Evidence, supported by mathematical models, indicates that the eruptions are apparently caused as a result of magnetic activity, like that which causes solar flares, but at intensities that are tens of times higher.

Solar flares are the result of a sudden release of magnetic energy accumulated in active regions of the Sun's atmosphere, where sunspots appear. The magnetic energy that is released causes charged particles to accelerate to velocities approaching the speed of light, and these collide with the solar gas and emit strong radiation (that looks like a flame). Some of the accelerated particles reach Earth, where their influence is discernable. For instance, the Northern Lights are caused by the collision of these charged particles with air molecules in Earth's atmosphere. When these particles interact with the Earth's magnetic field, they cause a magnetic storm that can paralyze satellites and communication devices (as indeed happened in some cases). Great efforts are, therefore, being invested in the development of means for early detection of solar flares. [1]

How is it possible that we can witness magnetic flares on magnetars but not on ordinary pulsars? In Chapter 11, we described the layered structure of the neutron star. We explained that the outer part of the star is composed of a kind of solid crust, about a kilometer and a half thick. The magnetic field lines of the star penetrate this crust, from the center out. Since the crust is a nearly perfect conductor, the field lines are frozen in the crust and cannot move, unlike what happens in the Sun. In ordinary pulsars, this state is permanent, and so there is no magnetic activity on

---

[1] It usually takes the energized particles several days from the moment of flaring until they reach Earth.

these stars that could cause eruptions like the ones we see on the Sun. Magnetars, on the other hand, have a magnetic field that is so strong that it can rupture the solid crust. When magnetic activity occurs, the field lines become twisted and distort the outer crust to such an extent that it shakes and ultimately fractures in certain places. This process ends in a huge release of the accumulated magnetic energy, which leads to the observed gamma ray flare. The intensity of the flare varies. Like earthquakes, most flaring events are relatively weak. Occasionally there is a strong one, but these are rarer. Even less frequently, the magnetic field undergoes a large-scale rearrangement, and then very powerful gamma ray burst may occur, like the one documented in 2004. During such an eruption, magnetic energy turns into hot plasma and radiation, mostly high-energy emissions that we can see. The tremendous pressure that is created as a result causes matter to be thrust off the surface of the star at high velocities. A few objects presented evidence of these fast matter jets. Nonetheless, because of the powerful magnetic fields, part of the plasma created in the eruption, which contains mainly electrons and positrons, remains trapped near the surface of the star in a kind of magnetic bubbles or loops, like those seen on the face of the Sun during solar flares (see Figure 1). The trapped gas rotates along with the entire star and emits gamma rays in the process. When the eruption is strong enough, this emission is visible for a very long time after the prompt phase (see Figure 2). Its intensity is much smaller than that of the radiation emitted during the prompt burst (which lasts less than 1 sec), and it diminishes over time with the cooling of the trapped gas.

The active phase of a magnetar is short compared with that of an ordinary pulsar. The magnetic fields slow down the magnetar's rotation rate, like they do in ordinary pulsars, but because the magnetic fields are so tremendously strong, the spin down time of magnetars is a million times shorter than ordinary pulsars. A typical magnetar loses most of its rotational energy within several dozens of seconds after its birth. In addition, the star emits a large amount of energy following the magnetic discharges described above, both in instantaneous explosions and in the incessant emission of X-rays, gamma rays, and particular matter. These processes cause the magnetic field to decay after approximately 10,000 years and

Figure 1.   A solar flare caught in action by the TRACE satellite. In the center, a magnetic (coronal) loop is seen leaving the face of the Sun. Courtesy of TRACE, a mission of the Stanford-Lockheed Institute for Space Research, and part of the NASA Small Explorer program.

subsequently to stop emitting X-rays and gamma rays, at which time the magnetar becomes inactive. The age of most magnetars observed so far, as estimated from measurements of their magnetic field strength and spin down time, is several thousands of years.

We mentioned previously that the magnetic field strength of an ordinary pulsar might be explained by the compression of the magnetic field of the core of the progenitor star. This process cannot, however, account for the powerful magnetic field measured in magnetars. What therefore is the source of the magnetic field of magnetars? According to current theory, magnetars have internal active dynamos, similar to the one that produces the magnetic field on Earth. The dynamo process kicks in when the rotation rate of the neutron star created after the collapse of the progenitor star exceeds a certain value. Neutron stars whose rotation rate does not

Figure 2. Illustration of a magnetic discharge on a magnetar: the tremendous magnetic pressure within the star ruptures the solid crust that envelopes it and enables magnetic field loops to burst out, subsequently causing a huge release of energy, some of which becomes the radiation that we see. Illustration: NASA/CXC/M.Weiss.

reach this threshold become ordinary pulsars. In the rare cases in which the rotation rate exceeds this value, the dynamo will enhance the original magnetic field, and the result will be a magnetar. It is estimated that a rotation rate of at least 100 rev/sec is required for a magnetar to be created. For the sake of comparison, the rotation rate of the Crab Pulsar at birth was about 50 rev/sec, too slow to become a magnetar.

Chapter 15

Black Holes — The Crown Jewel of Einstein's Theory

Since the possible existence of black holes was predicted in the early 20th century, black holes have captured human imagination. This mysterious object has been the inspiration for dozens of science fiction books and movies and for countless television programs. Hyperspace leaps used by protagonists in Star Wars and in other science fiction series, time travel, and even just an endless, unrestrained fall into a galactic hole — all of these phenomena are based on the physics of black holes. Black holes are also associated with information theory and with the possibility that the universe in which we live is a kind of holographic image. So what then is a black hole?

1. Matter, Energy, and Spacetime Distortions

According to the theory of general relativity, the gravitational force we feel in our daily life is the result of distortions in the spacetime fabric. In the absence of matter and energy, the spacetime would be flat,[1] and could be likened to a flat piece of paper. When a mass is present in this space, it bends the spacetime around it (Figure 1). This bending diverts the trajectory of a second mass that is moving in the same space, which we, the observers, interpret as if they were caused by a force of attraction between the two masses. Thus, for instance, the revolution of Earth

---

[1] Some curved spacetime manifolds do not contain matter and energy; these solutions contain a quantity called the cosmological constant, as explained in Chapter 1.

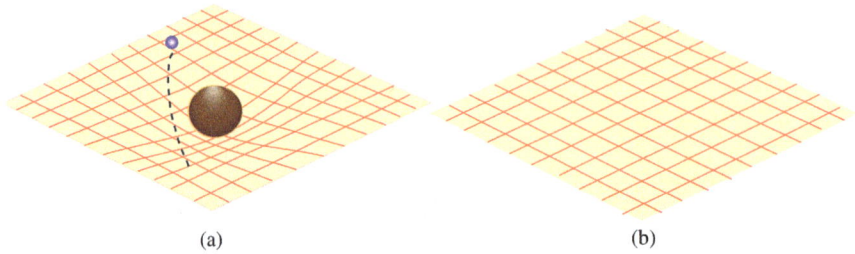

(a)   (b)

Figure 1. The smooth plane in (b) represents a flat spacetime, devoid of matter and energy. When this spacetime contains a mass, as illustrated in (a), distortions are created in its vicinity that affect the motion of other bodies. For instance, the smaller mass (blue ball) will travel along a curved trajectory represented by the dashed line.

around the Sun results from the spacetime distortions caused by the Sun. As mentioned earlier, the quantitative relationship between the distribution of matter throughout the spacetime and the curvature of that same spacetime is described by Einstein's field equations.

The denser the mass, the greater the distortions, until ultimately, the fabric rips when the mass collapses into a black hole. The black hole can, therefore, be likened to a tear or to a bottomless drain that is created in the spacetime. A mass that is trapped in the matter that is swirling around near a black hole, will end up being sucked in and disappearing from space, like water going down the sink drain.

2. Singularity and the Event Horizon — A Unidirectional Membrane

Mathematically, a black hole is a solution to the Einstein field equations. An in-depth understanding of the theory of black holes requires detailed understanding of the theory of general relativity, which is beyond the scope of this book. Nevertheless, the main characteristics of these objects may be described in a relatively simple manner. The simplest structure is a spherical black hole, which is characterized by only one parameter — its mass (see Figure 2(b)). It is called *spherical* because this solution has the symmetry of a perfect sphere. The spherical black hole solution was first derived by a physicist named Carl Schwarzschild in 1916. The radius of the black hole, which is called the *Schwarzschild radius* and is commonly

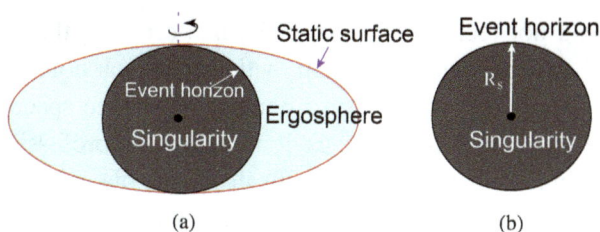

Figure 2. (a) Rotating block hole. (b) Spherical black hole.

denoted R_s, increases with the increase in the black hole's mass. To be precise, a black hole's radius is proportional to its mass. The Schwarzschild radius of a mass that is equal to the mass of our Sun is about 3 km. Since the radius of the Sun is approximately 1,000,000 km, we would have to shrink the Sun to about three millionths of its present size in order to turn it into a black hole. In that case, the density of the Sun would increase to 30,000 trillion times or 3×10^{16} its current density. Such compression does indeed take place when the core of an ordinary star collapses after its fuel is exhausted.

The spherical surface of radius R_s, which is called the *event horizon*, separates the interior of the black hole from its external environment. At large distances from the event horizon, the gravitational field of the black hole is, to a good approximation, similar to that of an ordinary star with the same mass and may be described using Newtonian theory. The closer you get to the event horizon, the stronger the gravitational field becomes, and the more prominent the effects of general relativity become. The event horizon acts as a unidirectional membrane: any physical object whatsoever, including any kind of radiation, can only move inward. No force in nature can make an object that is approaching the event horizon change direction and escape to infinity. This is also why they are called a "black hole" and an "event horizon". In fact, the above description is not completely accurate. The quantum theory allows black holes to "evaporate" in a process named *Hawking radiation*, after Stephen Hawking who first discovered the phenomenon. However, the heavier the black hole, the weaker the radiation, and so for the black holes discussed in this book, it is absolutely negligible.

Despite the strangeness of the event horizon, the forces acting on a free-falling body when it traverses the horizon are finite. In fact, a free-falling observer would not feel anything special when crossing the horizon. The peculiarities begin beyond the horizon, within the black hole. At the center of the black hole is the *singularity*, a point at which the spacetime distortions, or the *Riemann curvature* as it is called in professional jargon, increase to infinity. What exactly happens inside the black hole? What is its internal structure? What physics governs it? These questions do not yet have definite answers, and they are the subject of some ongoing, fascinating research. The strange phenomena that take place within the black hole, behind the event horizon that conceals its interior from the sight of the external universe, have inspired some of the descriptions in science fiction books and movies. Later on we will review some of the exotic phenomena related to the spacetime properties of the interior of the black hole.

3. Rotation and Energy Pumps

Another solution to the Einstein field equations describes a rotating black hole (see Figure 2(a)). This solution was discovered only in 1963 by a researcher by the name of Roy Kerr from New Zealand and is much more complicated than Schwarzschild's solution. The rotating hole is characterized by two parameters: its mass, like in the case of the spherical hole, and its angular momentum, which is related to the rotation. In fact, it can be shown that the spherical black hole is a special case of the Kerr solution, in which the angular momentum is zero. The rotating black hole has two characteristic surfaces, an inner surface and an outer surface. The inner surface is the event horizon through which nothing can exit. The outer surface is called the *static surface*. The region enclosed between the event horizon and the static surface is called the *ergosphere*. Objects in this region can stay in it without falling into the black hole, but they cannot remain at rest relative to a distant observer; they must rotate. In fact, the rotation of the black hole "drags" with it the entire spacetime that is outside of the event horizon, just like stirring a teaspoon around in a cup of tea causes the entire body of liquid to swirl. This dragging of the spacetime, commonly referred to as *frame dragging*,[2] forces bodies, wherever

---

[2]This effect was first described by Josef Lense and Hans Thirring in 1918 and is also known as the Lense–Thirring Effect.

they are, to circle around the black hole and in order to oppose this rotational coercion, such an object must apply a counter force (a rocket motor, for instance). The effect is very weak when the object is at a distance from the black hole, but gets stronger the closer you get to the event horizon, and within the ergosphere it is so strong that there is no force in nature that can resist it.[3] Thus, any body that enters the ergosphere is doomed to keep circling there forever. This effect exists only in the theory of general relativity, and it has no parallel in Newtonian theory, with which we are all familiar.

The frame dragging effect is caused not only by black holes, but also by any rotating object with a mass, and especially Earth. Calculations have shown that a satellite circling Earth will feel this effect, but its influence will be so small that it may not be measurable using ordinary means. In 1964, NASA initiated a project called Gravity Probe B, whose main objective was to develop a technology that would enable such measurement. In 2004, after 40 years of research and development, the experiment was launched into space on a satellite orbiting at a height of 642 km above the face of the Earth. The experiment consisted of four gyroscopes that were a million times more accurate than the best navigation gyroscopes available. The frame dragging effect was supposed to cause the axis of each of the gyroscopes to deviate (an effect known as precession) in a pre-calculated direction and to a pre-calculated extent.[4] The development of the gyroscopes was in fact the key to the success of the experiment, and constituted one of the greatest technological challenges that science has ever faced. A special telescope that was installed on the satellite made sure that the system's axis was directed towards a given point so that the deviation could be measured with the required accuracy. Data were collected for approximately one year. It took several more years to analyze them, and in May 2011 they were first published. The measurements matched the predictions, and one of the most important verifications of the theory of

---

[3] The physical reason for this is that in order to appear stationary to a distant observer, the object must move faster than the speed of light with respect to the local spacetime, which is forbidden.

[4] Precession is what causes the rotation axis of a top that is revolving at a high speed to rotate in a secondary circle before it topples over. But the reason for precession in this case, of course, is not frame dragging.

general relativity was thus provided. It seems that Einstein was correct — gravity does indeed reflect the geometry of the spacetime.

A short time after Roy Kerr published the rotating black hole solution, a British scientist named Roger Penrose made a dramatic discovery. He showed that within the ergosphere there are negative energy orbits. Penrose even described a thought experiment in which a particle that enters the ergosphere decays into two new particles (such processes occur routinely in nature), so that one particle falls into the black hole through a negative-energy orbit, and the other particle escapes to infinity. The particle that falls into the black hole adds negative energy to it, or in other words, causes it to lose energy. The energy that the black hole loses is gained by the particle that escapes to infinity, as is required by the law of conservation of energy. Thus, it is in fact possible to "suck" the rotational energy of a rotating black hole. The maximum amount of energy that can be extracted in this manner is about one-third of the total initial energy of the black hole. It soon turned out that under natural conditions, the energy extraction mechanism that Penrose proposed is not especially efficient. In 1977, however, physicists Roger Blandford and Roman Znajek showed that energy may indeed be extracted very efficiently using magnetic fields that penetrate the ergosphere. These magnetic fields are created in matter that is located in the close vicinity of the black hole. It can be shown that, theoretically, this is the most efficient way possible to produce energy in the universe. Many scientists today believe that this process is indeed responsible for the creation of jets of magnetized matter that travel at velocities that closely approach the speed of light, which were discovered in many systems throughout the universe, as will be elaborated on in Chapter 16.

The amount of energy that may be extracted from a rotating black hole is monumental. In order to get a sense of this, consider that the amount of available energy stored in a black hole whose radius measures a tenth of a nanometer, in other words, the size of a hydrogen atom, equals the amount of energy the Sun provides Earth with over a billion years. Such a microscopic device could, in principle, fulfill the needs of humanity for tens of billions of years. But before we get too excited, we must remember that despite its tiny size, such a black hole weighs 100 trillion tons, so that we cannot simply install one of them in our cellphone instead of a battery pack.

4. Time Stood Still

We already mentioned that the presence of mass and energy affects the geometry of spacetime. The immediate consequence of this is that the flow of time, as measured by a given clock, depends on the gravitational field at the point where the clock is located. To illustrate this, imagine two clocks that are positioned at different distances from a black hole, and which are at rest relative to a very remote observer (see Figure 3). The time measured by the observer at Clock A will flow much slower than the time measured by the observer at Clock B.

In general, every gravitational field, regardless of its source, has a similar effect on time, but the weaker the field, the smaller the effect. Earth exerts a similar effect on clocks that are mounted on satellites orbiting Earth. Although the effect is very small in this case, it is measurable. In fact, this effect is taken into consideration in the advanced GPS systems that are in wide use today, in which especially high precision is required. The movement of time becomes markedly slow in the vicinity of black holes due to their tremendous gravitational fields.

An obvious question is: How much time will pass, according to an observer's clock, until a space probe launched in the vicinity of a black hole reaches the event horizon? Detailed calculations show that for a spherical black hole, the time required is endless. The reason for this is that the closer the space probe gets to the black hole, the slower time, as measured by the distant observer, passes until it comes to a stop on the event horizon itself. It is even more interesting to note that the time, as measured by a clock located within the space probe itself, will be absolutely finite. As Einstein said, everything is relative. This fact raises doubts regarding the possible creation of black holes. If a black hole is created as a result of the death and collapse of an ordinary star in the galaxy, as described above,

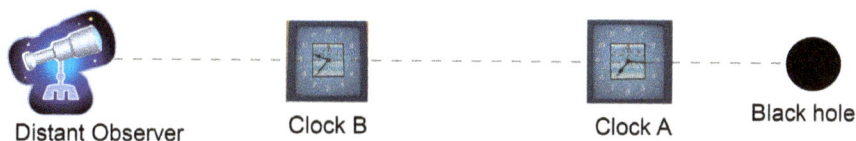

Distant Observer   Clock B   Clock A   Black hole

Figure 3. Two clocks at different distances from a black hole measure time differently. The time measured by Clock A will flow slower than the time measured by Clock B.

then the time that will pass on the clock of an observer on Earth until the core of the star shrinks to the dimensions of the horizon is endless. In other words, the time it takes to create a black hole in this process is seemingly longer than the age of the universe itself. If so, are black holes even created in the universe? And if the answer is yes, then how? Many researchers were, and still are, concerned with this question, whose answer is embedded in the concept of time itself.

5. The Smallest Black Holes

Can any body, small as it may be, be compressed to create a black hole? The answer is no. The smallest black hole that can possibly be created has a mass of about 20 millionth of a gram (to be precise, 21.7645 millionth of a gram). This mass, which is called *Planck mass* after the German physicist Max Planck who was mentioned earlier, is much larger than the mass of atoms and molecules, and so it follows that black holes cannot be created from chunks of matter that are too small. The radius of a Planck mass black hole is very small — only 1.61×10^{-33} cm. This length unit is called *Planck length*. For the sake of comparison, consider that the radius of a hydrogen atom is approximately 10^{-8} cm. This means that Planck length is 10 trillion times smaller than an atom. The tiniest black hole possible in nature is one whose radius is equal to Planck length. Where does this limit come from?

The limit on the mass of a black hole originates in quantum theory. As explained earlier in Chapter 4, according to quantum mechanics, every natural system has a kind of split personality — it behaves both as a particle and as a wave. The wavelike nature of a particle reflects quantum deviations from its classical trajectory, as predicted by Newton's laws of mechanics or, more precisely, the probability that a particle is in a certain place in a given measurement. These quantum deviations are of the order of the de Broglie wavelength of the particle and depend on its mass — the greater the mass, the smaller the quantum deviations. In the case of a black hole, as long as its mass is large enough, its wavelength is smaller than its radius, and the quantum deviations occur within the event horizon and have no effect on the region outside it. But for masses that are too small, the wavelength is greater than the radius of the event horizon, and so the quantum deviations prevent its creation. In other words, the moment we try to create a black hole from matter with too small a mass, there will always be part of the matter that is suddenly be outside of the black hole due to those quantum deviations, and this will prevent the possibility of containing all of the matter simultaneously within the event horizon. The mass for which the wavelength of a black hole is exactly equal to its radius, R_s, is Plank mass. The wavelength of a black hole of this mass is exactly equal to the Plank length. For masses that are greater than the Plank mass, the wavelength is smaller than the radius of the event horizon, and its existence is possible. For masses that are smaller than Plank mass, the quantum deviations are too large, and the existence of the black hole is impossible.

6. Black Holes Have No Hair

So far, the black holes we've described were characterized by two parameters: mass and angular momentum. A question immediately rises: Is it possible that there are additional parameters that will generalize the solutions describing black holes? To gain some insight, consider the following thought experiment: Suppose that electrically charged particles are added to a black hole that is isolated in the universe. For example, let electrons that are in the vicinity be swallowed beyond the event horizon. What then would be the effect of the added charges on the black hole? If the charge conservation law is valid, as physicists believe it is, then we would expect an electric field to appear around the black hole. Indeed, there are solutions for a charged black hole, and they include an electric field in addition to the gravitational field. And would a magnetic field also appear if the black hole is rotating? And in general, could we imagine the appearance of additional fields that are related to other conservation laws? Well, there is a conjecture according to which every solution of the Einstein field equations that describes a black hole is characterized by only three parameters: mass, electric charge, and angular momentum. This hypothesis in known as the no-hair theorem and it means that an observer outside of the event horizon has very limited information about the content of the black hole. For instance, it is impossible to distinguish between a black hole that is made of matter and a black hole that is made of anti-matter if they both have the same mass, charge, and angular momentum. Indeed, in general, there is no possibility of knowing for certain what is inside a black hole. No detailed information can cross the event horizon and escape into the external world. The sole remnants of a piece of matter that has disappeared forever into the depth of the black hole are its energy, its electric charge, and its angular momentum, which are added to the black hole. Systems that include black holes fulfil the laws of conservation of mass, charge, and angular momentum, as required of any physical system, but it seems as if they are losing information.

7. Black Holes and Information Bounds

What is the maximum amount of information that can be stored in a given volume? This question has intrigued, and still intrigues, many researchers in the world of science and technology since the mid-20th century. The first computers developed in the 1950s were the size of a room and could perform only the most basic arithmetic calculations, which today may be performed using a microscopic device. The amount of information that can be stored on a personal computer's hard drive has increased in the past two decades almost 1,000-fold, and this growth rate continues also today. The main reason for this minimization ability is that new ways are being found to code the basic units of information — bits — in tinier and tinier elements. Is there a theoretical limit to this minimization process? And if yes, what is it? The answer, as we will see shortly, involves black holes.

We know from experience that a glass bottle that falls from a table to the floor will shatter into pieces. Nobody has ever seen fragmented glass jump back up on to a table and reassemble into a bottle. When ice is placed in hot water, the ice melts. Nobody has ever seen a situation in which the ice gets colder and the water gets hotter. A more fragrant example is a bottle of perfume. When we open a bottle of perfume in the middle of a room, the perfume in the bottle will begin to evaporate, and after a while the entire room will be filled with its scent. Is the reverse process possible? In other words, can we disperse perfume throughout a room, and then have the perfume collect back into an empty bottle that is sitting in the middle of the room? Experience teaches us that it is impossible. These are just three examples of a huge number of processes that are called *irreversible.* The bottle can shatter to pieces, but pieces will never become a bottle. Heat always flows spontaneously from the warmer body to the colder body and never in the opposite direction. Perfume will always evaporate and disperse and will never collect. Why is that? The flow of heat from a cold body to a warm body and the creation of a bottle from glass fragments are not in defiance of the law of conservation of energy. What law of physics then prohibits it?

Recurring attempts have been made throughout history to build perpetual motion machines, trials that began in the 8th century and turned into an obsession in the 19th century. The idea was to build a device that after being activated once would continue in its movement forever and would never stop. In other words, no spring would have to be wound, no fuel would have to be refilled, nor would it be necessary to connect it to any other external source of energy. Some of these contraptions contradicted the law of conservation of energy, but others, mainly those that spontaneously turn heat into work, did not contradict the law of conservation of energy, and it seemed quite possible that they would be realized. The industrial revolution that began at the onset of the 19th century with the development of steam engines, prompted engineers and physicists to thoroughly investigate heat and its properties and to try to understand the theoretical limitations on the efficiency of the engine and of the perpetual motion machine, with the objective of building more efficient engines. These investigations ultimately led to the definition of a quantity called *entropy*, which is a measure of the disorder of a physical system, as will be explained below, and to the formulation of one of the most important laws of physics — *the second law of thermodynamics* — that states that the entropy of any closed system, that is a system that is completely isolated from its environment, can never decrease.[5] This law is what prevents fragmented glass from becoming a bottle, and heat from flowing from a cold body to a hot body. The second law does not, however, prevent the diminishing of entropy in a system that is not isolated. For instance, a refrigerator is a device that transfers heat from a cold body (the interior of

---

[5] The first law is the *law of conservation of energy*. There is also a third law, whose simple formulation stipulates that the entropy of a system whose temperature approaches absolute Zero, also approaches zero.

the refrigerator) to a hotter body (the air outside the refrigerator). The entropy inside the refrigerator decreases, but outside the refrigerator it increases, since the refrigerator motor that activates the compressor heats the surrounding air. The increase in entropy outside the refrigerator exceeds the loss of entropy within it, so that the total entropy of the system increases, as required by the second law. In fact, cooling the inner part of the refrigerator, and thus decreasing the entropy there, requires the transfer of heat from the inside out (and so the interior of the refrigerator cannot be isolated). This transfer of heat leads to an increase in entropy outside the refrigerator.

In 1877, Austrian physicist Ludwig Boltzmann showed that entropy could be related to the number of microscopic states which the elements that make up a system can occupy without changing its macroscopic state.[6] To illustrate what this means, imagine a system made up of the combination of four elements, each of which can receive two values, 0 or 1. A given arrangement of the four digits will be called a microscopic state. For instance, 1001, 0001, and 0110 are possible microscopic states of the system. It is obvious that there are a total of 16 different microscopic states. Now, let's assume we have a device that displays the sum of the digits of a given combination that is input into it, and let's define a macroscopic state as a specific sum of digits. For example, the device will display the value 2 when the combination 1001 is input into it since $1 + 0 + 0 + 1 = 2$, while the combination 1110 will result in the value 3. It is emphasized that the device cannot distinguish the microscopic state of the system (that is, the arrangement of the digits), but can only calculate the sum of the digits. Let's suppose that an unknown combination of digits was input, and the device displayed the value 0. We know in absolute certainty that when the device records the macroscopic state 0, the system must be in the microscopic state of 0000; there is no other combination whose sum of digits is 0. On the other hand, if the device displays the value 2, we know that the system must be in one of the following six microscopic states: 0011, 0101, 0110, 1001, 1010, or 1100, but we do not know which. It may be said that in the macroscopic state 2 there is greater

---

[6] To be more precise, for equally probable states, the entropy is the natural logarithm of the number of possible microscopic states in a given macroscopic state.

disorder, or greater entropy, than in the macroscopic state 0, since there are more possibilities of obtaining this state. If we were to arrange the digits in a random manner, the system would have a higher probability (six times higher) of being in state 2 than in state 0. Hence, the system has a higher probability of being in a state of higher entropy. The reason natural systems tend to be in states in which entropy is as high as possible is that the probability to be in such a state is maximal.

To further demonstrate the concept of entropy, imagine what happens during a billiard game. At the beginning of the game, the target balls are arranged in perfect order on one side of the table, while the white ball is placed at the other side. There may be more than one way of arranging the balls before the opening strike (depending on the kind of game played), but in any case the number of possibilities is very limited. In this state, the system's entropy is low. After the breaking strike, the balls scatter every which way at different speeds. It is clear to us all that the disorder is now greater. It is also clear that the number of possibilities the balls have to scatter is immense. We can say with very high certainty, that after the breaking strike, the balls are scattered randomly, and are at greater distances from one another than at the beginning, but we are unable to predict the exact arrangement of the balls after the breaking strike. If we were to repeat the process a large number of times, we would discover that the balls are arranged completely different each time. The chance of obtaining the same (or even similar) arrangement of the balls twice after the breaking strike is negligible — even if we were to repeat the process millions of times or even billions of times, we would probably never succeed in repeating the same result. In this state the entropy, which represents the number of different states the target balls can be in after the breaking strike, is much higher than in the initial state. The entropy of the billiard ball system increases in the time that passes from the starting point, slightly before the breaking strike until the balls stop moving following that strike.

The same principle applies to any physical system. Take for example, the air in a room: a certain temperature and pressure define a given macroscopic state of the air. On the microscopic level, air is composed of a huge number of molecules that move about randomly at various velocities. Measuring the temperature of the air using a thermometer reflects the average velocity (or more precisely, the average kinetic energy) of the air molecules. From reading the temperature alone, it is impossible to know how fast each molecule is moving, or even the number of molecules moving at a certain velocity. In other words, we would not be able to know if someone were to change the arrangement of the molecules in the room without changing the air's temperature and density. There is a tremendous number of ways in which the molecules can move so that the same temperature is obtained. Each of these ways defines a microscopic state. Entropy constitutes a measure of the total of all possible microscopic states for a given temperature. Or more precisely, the entropy of the system is directly proportional to the logarithm of the number of these microscopic states.

The fact that the entropy of a physical system must increase is most deeply connected to time's property of moving in one direction — from the past to the future. We will not delve into a discussion in this context, but we will mention only that the fact that the universe is expanding, that we get old, and that stars die is the result of entropy increasing.

So what is the connection between entropy and information? In 1948, an American mathematician named Claude Shannon published a seminal article in which he formulated the principles of information theory. In his paper, Shannon showed that the amount of information contained in any message is equivalent to entropy, as defined by Boltzmann. If we return to the digit system example presented earlier, in the 0 state, no information may be coded, because there is only one option of forming this state — 0000. In the 2 state, certain information may be coded already: for instance, we could decide that the series 0011 is equivalent to the letter A, the series 0101 to the letter B and so on. According to this method, we could create a message with up to six English letters. There are, of course, more efficient ways of doing this. Loosely speaking, the entropy of a message, according to Shannon, is the minimum number of bits required to encode the message. According to this definition, the entropy of state 0 is 0 and the entropy of state 2 is about 2.586, which is the logarithm to the base of 2 (or the binary logarithm) of the number of microscopic states, which is 6. Knowing the entropy alone does not tell us a thing about the content of the message, but it serves as a measure of the amount of information. The design of any modern means of communication — from mobile phones to CD players — and the methods of compressing information on the internet all rely on Shannon's theory.

It is important to emphasize that Shannon's entropy and Boltzmann's entropy are not identical for a given system. Each memory component in a silicon chip has two states: 0 (off) or 1 (on). On the other hand, such a component is made up of a huge number of atoms, and a large number of possible atomic arrangements will yield the 0 state or the 1 state. It therefore follows that Boltzmann's entropy is much greater than Shannon's entropy. The smaller the memory component, that is the fewer atoms it is made up of, the closer the two entropy values. Increasing minimization is bringing nearer the day when every atom will store one bit of information for human usage, and so Shannon's entropy of the most advanced chips is destined to gradually approach the order of magnitude of Boltzmann's entropy of the chip compounds.

A serious problem arose upon the discovery of black holes. As mentioned above, the no-hair theorem stipulates that a black hole is characterized by three parameters only: mass, charge, and angular momentum. The information about the exact composition of the mass or charge is hidden from an outside observer by the event horizon. Information that has penetrated into the black hole through the event horizon is seemingly lost or at least "concealed" from the eyes of the outside world. The immediate meaning of this is that in the formation process of a black hole, the system's entropy decreases, an occurrence that contradicts the second law of thermodynamics. In 1972, Jacob Bekenstein, a young student at Princeton University (and later professor of physics at the Hebrew University of Jerusalem until his recent death in 2015), proposed an elegant solution to this problem. Bekenstein claimed that the entropy of a black hole is directly proportional to the surface area of its event horizon, or more precisely to one-fourth the area of the event horizon,[7] measured in Plank area units (see Figure 4). Bekenstein also asserted that the second law of thermodynamics applies to the entropy of black holes like it does to all physical systems that were known up until that time. This claim was based on works conducted two years prior, in which researchers showed that in various processes such as in the merging of black holes, the total area of the event horizon never decreases.

---

[7] The factor ¼ in the expression for the entropy was derived later by Stephen Hawking.

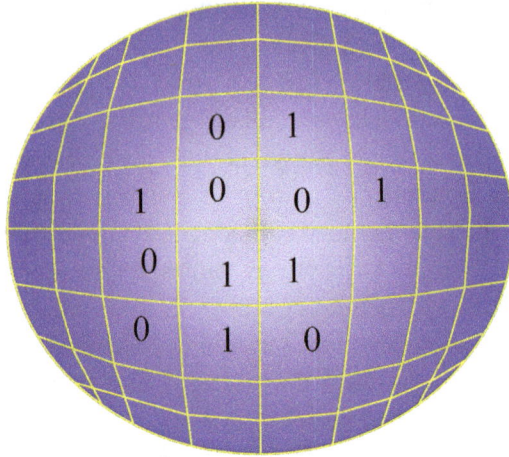

Figure 4. The entropy of a black hole is the number of squares that cover the event horizon, where the side of each square is equal to Planck's length, 1.61×10^{-33} cm. Each square may be regarded as one bit of information with two states, 0 and 1. For a black hole with a 1 cm radius, there are approximately 10^{66} such bits, which represent the entropy of the black hole.

This, however, did not entirely solve the problem. About three years after Bekenstein published his proposal, Stephen Hawking showed that black holes are not completely black, but rather evaporate while emitting radiation referred to as Hawking radiation. How is this possible? Indeed, we said that nothing can escape from within a black hole. Well, the answer lies in the quantum theory. The same quantum uncertainty, which we already encountered several times in the previous chapters, is what enables the emission of particles and radiation from a black hole into the universe outside of it. The evaporation rate of a black hole is smaller the greater its mass. For a black hole with a mass equal to the mass of our Sun, this evaporation time is much longer than the age of the universe. But for very small black holes, this time may be sufficiently short. If so, according to Hawking, a sufficiently small black hole that was somehow created will eventually evaporate and turn into a collection of measureable radiation and particles. Hawking also showed that the radiation is emitted from a small region near the event horizon and therefore cannot contain the information that is located in the interior of the black hole. Hence, information is lost in the process in which a black hole is created and evaporates. For example, let's suppose that we created two black holes. One was created by compressing a freight train, while the other was created by compressing a residential building with the same mass as the train. It is obvious that before the black holes were created, we could easily differentiate between the train and the building, that is, each of the systems possessed different information. After creating the black holes and after their evaporation, it is impossible to know, just from measuring the radiation, whether it came from the train or from the building. This apparent loss of information contradicts the *unitarity principle* of quantum mechanics that requires the information of pure states to be conserved. This contradiction, which was named the *black hole information paradox*, is the main reason that Hawking's discovery was controversial for close to 30 years. In recent years, Hawking has backed down from his position and admitted that his original theory requires some modifications. A breakthrough came when the *holographic principle* was developed, as will be explained later on, but there are still some fundamental issues to be dealt with, and it will take some time before the black hole information paradox is resolved. The current conventional wisdom is that the information of a piece of matter that has fallen into a black hole is embedded on the event horizon together with a large amount of "noise".

One of the most interesting conclusions that stem from Bekenstein's proposal regarding the entropy of black holes is that there is a limit to the amount of entropy (or information) that may be contained in a given volume or alternatively in a device with a given mass or energy. Let's suppose that we have stored information in a system that is not a black hole. We will now cause the system to collapse, so that a black hole is created. Since the entropy cannot decrease during the collapse process, but can only increase, the entropy of the original system must be smaller than the entropy of the black hole formed as a result of its collapse. In other words, black holes set a theoretical maximum on information storage — it is impossible to store a larger amount of information than that which is contained on the surface of the event horizon of a black hole that is created from the sum of the information-containing units (for instance, the memory chips in our personal computer). From a practical point of view, this amount of information is huge relative to the amount of information that can be stored today in computer memories or in any other means; a device of 1 cm^3 can contain up to approximately 10^{66} bits of information, while in reality, a computer chip that size contains no more than 10^{12} (a trillion) bits. Nevertheless, this universal limit is of practical significance.

8. Is the World a Hologram?

The insight that the information that a black hole contains is located only on its boundary (the event horizon) led Dutch physicist and Nobel Prize winner Gerard't Hooft to propose the holographic principle, according to which all of the information contained in any volume in space is encoded on the surface area of that volume. A practical example of this, and one that has lent its name to the principle is the hologram images we are all familiar with. A hologram is a three-dimensional image created by passing a laser light through a two-dimensional photographic plate that contains a collection of spots (see Box 1). All of the information required to reconstruct the image of a three-dimensional object, complex as it may be, is embedded in these spots on the flat photographic plate.

Gerard't Hooft, followed by Leonard Susskind of Stanford University, refined the idea, and suggested that every physical system, including the entire universe, is a kind of hologram. That is, the full description of the system is encoded on its surface area. This proposal seemingly contradicts logic, because volume increases faster than the surface area that contains it. To elaborate the point, Bekenstein proposed, in an article he published in *Scientific American*,[8] the following thought experiment: Let us imagine a process in which we collect computer memory chips into a large pile. The number of chips, and so the information stored on them, increases in direct proportion to the volume of the pile. On the other hand, the maximum limit on information capacity (the holographic limit) increases "only" in direct proportion to the pile's surface area. Since the volume grows faster than the enclosing surface, the amount of information stored in the volume grows faster than the holographic limit, and so if we continue to add chips to the pile, we could eventually reach a situation in which it would be impossible to store all of the information in the chip pile on its surface area. There is

---

[8] Jacob Bekenstein, *Information in a Holographic Universe*, Scientific American 289(2), (2003).

Box 1.   What is Holography?

Holography is a technique that uses interference of light to create three-dimensional images. As opposed to ordinary photography, in which only the intensity of the light reflected from the photographed object is retained, in holography, in addition to the light intensity, information is retained also regarding the phase of the wave, which enables to reconstruct the true three-dimensional image of the holographed subject. An image taken using this method is called a hologram. Holography was invented by chance in 1947 by Hungarian physicist Dennis Gabor during his attempts to improve an electron microscope.

According to the original version of the invention, a photograph is taken using a laser beam that is split in two using a beam splitter. One beam strikes the photographic plate directly, while the other strikes the object being photographed and is reflected back from the object to the photographic plate. Thus, an interference pattern is obtained on the photographic plate, which contains all of the information regarding the three-dimensional object that was photographed. To the naked eye, this pattern looks like a random collection of dark and light spots. To reconstruct the image, a laser light is passed through the photographic plate at an appropriate angle (see below figure), and the object's three-dimensional image appears on the other side of the plate. As surprising as it may sound, all of the information contained in the three-dimensional body is in fact encoded in the spots on the two-dimensional photographic plate.

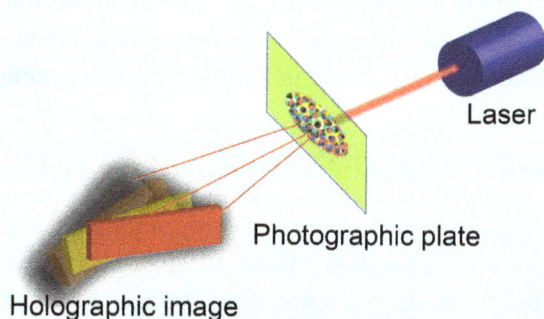

Laser

Photographic plate

Holographic image

seemingly a contradiction between the holographic principle and the second law of thermodynamics — either the entropy decreases when chips are added or the holographic principle is erroneous. In fact, claimed Bekenstein, what will happen is that the pile will collapse into a black hole because of its self weight before we reach the crossing point, so that there is no real contradiction. The laws of physics, in this case the theory of general relativity, masterfully protect themselves from such contradictions.

In its deeper meaning, the holographic principle posits that a certain physical theory that describes a world in a given dimension is equivalent to another physical theory that describes what is happening on the boundary of this world. One of the more important examples is the equivalence between five-dimensional gravitational theories, specifically the string theory, and the four-dimensional theory of a non-gravitational quantum field, which constitutes the boundary of the five-dimensional world. For instance, a black hole in a five-dimensional world is equivalent to radiation that is emitted in the four-dimensional boundary of this world. If these theories are correct, then creatures that live in one of these universes cannot know whether they are living in a five-dimensional world that is described by a string theory or in a four-dimensional world that is described by a quantum field theory.

And perhaps the universe in which we live is itself a hologram? Such speculations exist, according to which our life in this universe is no more than a reflection of what is happening on the universe horizon — that plane from which the initial rays of light arrived (see Chapter 1). What alternative set of laws applies there? The answer to that question is not yet known. If all this is true, then the world may be made up of information units, and matter and energy are in fact its secondary components.

9. Black Holes, White Holes, Wormholes, and Time Travel

"The command spaceship sailed about throughout the expanses of the universe on its way to another routine mission. Suddenly, the ship's captain sensed the beginning of something strange that was going on. The ship's clock was showing irregular time changes, and the space curvature meter was warning about a growing spatial warping. Behind the space ship, stars were beginning to disappear, and an eerie darkness spread forth. In front of the spaceship, stars appeared as glowing streaks of light, evidence of the warping of the space around the ship. Before any of the crew members had time to understand what was happening, the spaceship was swallowed up into the mouth of a local wormhole, and was sucked into another universe."

Our readers will no doubt recognize this excerpt from countless scenes taken from science fiction movies. Do these ideas have a scientific basis or are they complete fantasy? Surprising as it may sound, the scientific debate regarding the possibilities of parallel universes and wormholes began very early — shortly after the publication of Schwarzschild's solution. In fact, some of the ideas for the science fiction series, like *Sliders*, were contributed by physicists who are studying those issues.

In the introduction to this chapter, we claimed that a space that contains a black hole is divided into two regions that are separated from one another by the event horizon, a unidirectional membrane that enables the movement of bodies and the flow of information only from the outside in. This description is not in fact complete. From a mathematical viewpoint, the solution that Schwartzchild found describes four different regions in the space. In addition to the two regions discussed above, there is a third region, called a *white hole*, and a fourth region, which describes another infinite universe that is similar in its characteristics to the universe we are familiar with. The white hole, like the black hole, is confined by a unidirectional membrane called the *anti-event horizon*, through which any physical object can pass only from the inside out. The center of the white hole contains singularity, like the singularity of the black hole. In fact, a white hole is a black hole in which time flows in reverse — from the future to the past. No causal connection exists between the two universes that are outside the two holes, the white and the black; a viewer who wishes to pass from one universe to the other, will invariably find himself inside the black hole, from where you can only reach singularity — the point at which the spacetime curvature increases to infinity. Nevertheless, the black hole may be penetrated through each of the universes, so that any object from one universe may encounter an object from the other universe within the black hole. Such a meeting will be very brief, since each of the two objects must continue to move in its path toward the singularity, anticipating the moment at which the forces of the endless tide

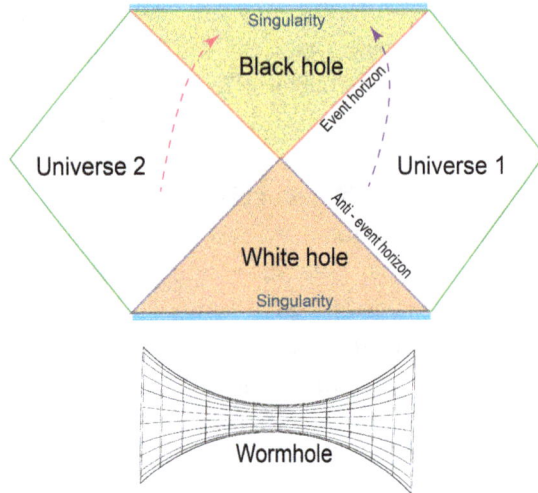

Figure 5. A spacetime diagram that depicts Schwarzchild's solution (top): The two universes are separated by a black hole and a white hole. Wormholes connect the two universes, which are entered through either the black hole's event horizon or the white hole's anti-event horizon. The trajectory of an observer who falls from one of the universes into the black hole must always end in the singularity and is described either by the purple line representing a fall from Universe 1 or by the red line representing a fall from Universe 2. Therefore, physical bodies are in fact prohibited from passing from one universe to the other through wormholes. The only place where occupants of the different universes may meet up for a very brief instance is within the black hole.

will tear it apart. Even though a physical object cannot cross from one universe to the other, the universes are interconnected through the black hole or the white hole. This linkage is called a *wormhole*. Figure 5 shows a spacetime diagram that depicts Schwartzchild's solution.

Do white holes actually exist in nature? Up to date there is no known way to create a white hole in nature, other than the possibility that white holes were created in the primordial universe through a process of some kind. Claims have been made according to which the creation process of a black hole itself, for instance due to the collapse of a star, is accompanied by the creation of a white hole and an additional universe, as the Schwartzchild solution enables, but not many actually believe so. Attempts have also been made to expand the Schwartzchild solution in different ways. One of the suggestions, which is an especially interesting one, involves the merging of the black hole's singularity with that of the white hole. The immediate implications of this is that an observer who falls into a black hole may exit through the white hole and meet himself in a different time. In other words, this solution enables time travel. This proposal was also rejected in its literal form, although it opened up the door to a more in-depth investigation of time machines.

Many years of research have led to the surprising conclusion that the Einstein field equations enable solutions that describe time machines and multiple universes. The mere fact that such a mathematical solution exists does not mean that it must actually materialize in nature. But if it is somehow possible to create the conditions required to realize the solution, then it may be realized. One of the ways in which physicists are trying to evaluate the feasibility of time machines and parallel universes is by assuming the existence of a given system, for instance, a certain kind of time machine, and then identifying the conditions required to realize the solution by applying the Einstein field equations. One important conclusion that researchers arrived at is that there is no physical law that prohibits the existence of time machines and multiple universes, but that restrictions to such time travel must exist so that the future does not affect the past. The second conclusion is that in order to realize such systems, a kind of exotic matter with negative energy is needed. It is possible, for instance, to expand and stabilize a wormhole by adding such matter to a given region in space so that physical bodies will be able to transition from one universe to a parallel universe. But where can such matter be found? Can it be created? Well, the dark energy of the universe is an example of this kind of matter.[9] Another example is the *quantum vacuum energy* that causes the *Casimir effect* — an effect predicted by Dutch physicist Hendrik Casimir in 1948. The Casimir effect describes the attraction that exists for no apparent reason between two plates that are situated in a vacuum environment. How can this kind of matter be produced and used to build time machines? This question is still without answer, and hundreds of years will probably go by before science acquires sufficient understanding in order to answer it.

---

[9] Dark energy has a positive energy density but negative pressure.

10. Space Warps and Space Leaps

Space leaps and especially warp drives are ways of traveling at speeds that exceed the speed of light, and are used in many science fiction series, such as *Star Trek*. Warp drive technology is based on creating a "bubble" of ordinary spacetime. Within this bubble is the spaceship, and the entire bubble sails around the warped spacetime in such a way that the bubble's motion creates a sense of velocity that is greater than the speed of light. Inside the bubble, the space curvature has no effect, and so no strong forces act upon those who are within the bubble.

The idea of curvature-based propulsion prompted Mexican physicist Miguel Alcubierre to see whether such a scenario is possible in the real world. This was, perhaps, the first time in which scientific research was influenced by a science fiction series. Alcubierre succeeded in showing that Einstein's field equations enable the existence of a strange space of that kind. According to his model, the space in the front part of the bubble shrinks, while the part in back of it expands, enabling the bubble to travel from place to place at a speed that exceeds the speed of light without contradicting the causality principle. Within the bubble itself, light still travels faster than anything else. And while huge gravitational forces resulting from the space curvature act at the edge of the bubble, within the bubble there is no acceleration and no forces act, so that everyone within the bubble is protected. The spaceship can be imagined to be riding a kind of wave (the bubble) in an external spacetime, like a surfer riding a wave in the sea on a surfboard. The surfer is unaware of any turbulence in the water beneath the surfboard, nor of the forces that the wave applies to the surfboard; she simply enjoys her surfboard ride on the water.

However, it is not simple to realize *Alcubierre's propulsion*, and many even claim it is impossible. First, Alcubierre showed that in order to create such a space, exotic matter with negative energy is needed. As we've already mentioned, there is evidence of the existence of such matter, so that this requirement does not disqualify the idea. But is it possible to accumulate such matter? And if so, how can it be used appropriately, if at all possible? These are questions of a different kind. Furthermore, a spaceship cannot simply enter such a bubble and start moving. Appropriate paths between the transition points must be prepared in advance, a kind of railroad tracks on which this bubble will be able to travel. When we talk about galactic-scale distances, this idea does not seem possible as for now. And there are other problems as well. Nevertheless, some believe that ideas of this kind will pave the way for future technology that will enable rapid travel throughout the solar system, and maybe even throughout the galaxy.

11. Extra Dimensions

A short time after the theory of general relativity was published, a German mathematician by the name of Theodor Kaluza asked himself the following question: What would happen if the universe we live in had another dimension? What form would Einstein's field equations take then? Kaluza was surprised by the answer, and he was not the only one. It took Einstein himself quite a while to digest it.

What Kaluza discovered was that in a world with five dimensions — four spatial dimensions and one time dimension — Einstein's field equations describe the electromagnetic force as well as the gravitational force. According to this theory, the electromagnetic force is the geometric reflection of the fifth dimension, just as the gravitational force reflects the geometry of the ordinary spacetime. This was the first time ever that someone had formulated a theory in which two of the familiar forces in nature were unified. No longer were they two fundamentally different forces, but rather two aspects of the same phenomenon — the geometry of a five-dimensional spacetime. The theory was expanded and enhanced by Klein, one of the leading physicists of the 20th century, and was subsequently called the *Kaluza–Klein theory*. The attempt to unify the four known natural forces into a single theory, whose seeds were sown by Kaluza and Klein, dictated the development of modern physics in the decades that passed since.

Our daily experience teaches us that there are always only three spatial dimensions. Where then is the additional dimension hiding? Klein's suggestion was that the fifth dimension (in addition to the fourth dimension, which is, of course, time) is circular, and that its radius is so small, much smaller than the size of an atom, that we cannot discern it directly. In fact, it can be shown that the property of the electromagnetic force

Figure 6. The red line on the left represents a unidimensional world. In this world, the only possible motion is back and forth along the red line. The right-hand part of the illustration demonstrates how this world changes when another closed dimension is added. The unidimensional line becomes the envelope of a cylinder. Motion in this world is possible in two directions, as the arrows indicate. In the direction of the blue arrow, it is possible to cover great distances. On the other hand, movement on the cylinder's envelope in the direction of the red arrow will lead us back to the starting point.

called *gauge freedom* requires that the fifth dimension be closed (circular). To understand what a closed dimension is, imagine a unidimensional world in which unidimensional creatures live (see Figure 6). The only thing these creatures can do is move back and forth along the line that constitutes their world. One day, a unidimensional physicist discovers that their world is actually a cylindrical surface of endless length, rather than a line as was believed up until then. It is still possible to move endless distances along the axis of the cylinder, but on the other hand, in the direction perpendicular to the cylinder's axis, movement is possible only in a circle; walking along the dimension perpendicular to the cylinder's axis will very quickly lead the walker back to the point he set out from. If the unidimensional creatures were smaller than the radius of the cylinder, they would probably have noticed that their world is in fact two-dimensional. But since the radius of the cylinder is much smaller than the size of these creatures, even smaller than the atoms they are made of, they were not able to distinguish the second dimension. Only through experiments conducted in the largest particle accelerators they possess — experiments that can delve into very small length scales — could they observe the existence of this dimension.

Most theories in modern physics that attempted to unify the gravitational force with the three other forces, like the string theory, require a world in which there are other, closed dimensions, in addition to the three ordinary spatial dimensions and the time dimension. The number of extra dimensions depends on the specific theory. The radius of the extra dimensions in most of the theories is equal to Planck's length, which is also the radius of the smallest black hole possible in nature, as explained earlier.

Energies that are hundreds of billions of times greater than the energies of the largest particle accelerators that exist today are needed in order to probe such scales, making it infeasible. Lately, however, new theories have emerged, in which the radii of the additional dimensions are much larger than Planck's length. If these theories are correct, it might be possible to probe the extra dimensions using the particle accelerators that currently exist, and especially using the Large Hadron Collider. Indeed, attempts are currently being made to conduct such experiments.

12. Particle Accelerators — Black Hole Factories?

To create a black hole, a large enough quantity of matter must be compressed into a radius that is smaller than Schwarzschild's radius. Particle accelerators, in which beams of particles collide with each other at very high energies, are the best way to compress matter artificially. Is it possible then to create black holes in particle accelerators? If the smallest mass needed in order to create a black hole is Planck's mass, as claimed above, then at least 20 millionths of a gram must be compressed in order to create a black hole. This quantity is tens of times greater than the mass that can be compressed in the Large Hadron Collider in a given time and so, in this case, it is impossible to create a black hole this way.

The new theories, on the other hand, in which additional dimensions exist that are larger than Planck's length, enable to create black holes with masses much smaller than Planck's mass. If these scenarios are correct, scientists might be able to create a black hole in the Large Hadron Collider in Switzerland. The creation of such a black hole will serve as irrefutable evidence of the existence of extra dimensions. It will even give a certain indication as to the number of those additional dimensions. Like in the above example, in which the unidimensional creatures suddenly discovered they were in fact living on the envelope of a cylinder, so will the creation of a black hole in a particle accelerator be a "penetration" into the extra dimensions that exist in our world, and will expose the "multidimensional manifold" on which we live.

The possibility that scientists may be able to create a black hole in the Large Hadron Collider was accompanied by much apprehension on the part of the general public. The main fear was that the black hole that would be created by the accelerator would begin to swallow matter and consume the entire Earth. These concerns have no scientific basis. First, such a black hole would evaporate immediately while emitting Hawking radiation. Second, cosmic rays that penetrate Earth have much greater energies than the energies that can be achieved in any particle accelerator. If these fears were justified, Earth would have ceased to exist a long time ago. Hence, there is no real risk in trying to create small black holes.

Chapter 16

The Black Holes' Magic Show

The rapid development of radio astronomy following World War II led to the discovery of many radio sources in the universe. The Third Cambridge Catalogue of Radio Sources, published in 1959, contained several hundreds of objects. The nature of these objects was unclear, and efforts were made to identify them with known sources of visible light discovered by ordinary telescopes; such efforts did not, however, meet with much success. The first identification took place in 1960, when a star-like object was discovered in the direction of one of the radio sources designated in the Cambridge catalog as 3C 48. To the researchers' surprise, however, the spectrum measured did not match that of the ordinary stars that were known then, and the nature of the source remained a mystery. In 1963, a breakthrough was made when an American astronomer of Dutch origin, Martin Schmidt, succeeded in identifying another radio source, 3C 273, in visible light, and measured its spectrum accurately using a telescope positioned on Palomar Mountain, California. Like in the first case, here too it turned out that the spectrum had different characteristics than those familiar theretofore. Schmidt did not give up, and after a systematic analysis of the spectral components, he reached the conclusion that it was not a new kind of matter, but rather a radiation source that was moving away from us at a stupendous velocity of 47,000 km/sec, about 15% of the speed of light. The irregular properties of the spectrum could be explained by the Doppler effect, which is caused by the motion of the source, and after taking this effect into consideration, Schmidt discovered that the light was being emitted from ordinary matter that primarily contained hydrogen. It was then suggested that the source was a cosmological source located very far from our Milky Way galaxy, and that its high speed actually stems from the expansion of the universe. Hubble's law led to the conclusion that the source was located about two billion light years away from Earth. Schmidt's discovery led to a reexamination of the previously discovered 3C 48 source, and indeed it turned out to be a manifestation of the same phenomenon.

The name these strange objects were given is *quasi stellar radio sources,* or *quasars* for short (see Figure 1). This name stuck and is still in use today. Later on, it was discovered that not all of the radio sources in the Third Cambridge Catalogue are quasars; many were identified, from an observational point of view, as various kinds of galaxies, and particularly as *radio galaxies*, which unlike regular galaxies are characterized by very bright radio emission, and as other objects, some of which will be addressed later on in this book. At the same time, a large number of quasars were discovered that did not emit radio waves. In fact, scientists found that only

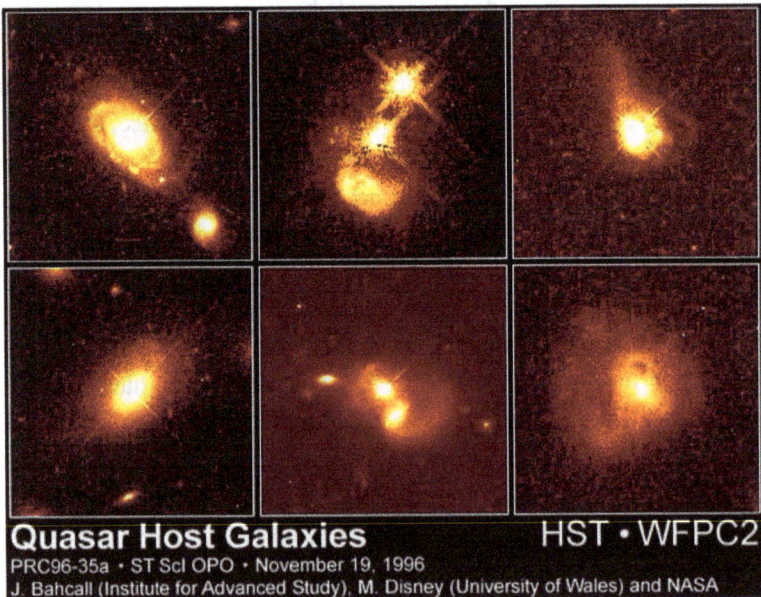

Figure 1. Examples of quasars. The bright spot in the center of each image is the light that is emitted from a small region near the black hole located at the center of the galaxy. The weaker light "smeared" around the bright spot is the light emitted by the stars in the galaxy. Courtesy of the Hubble Space Telescope.

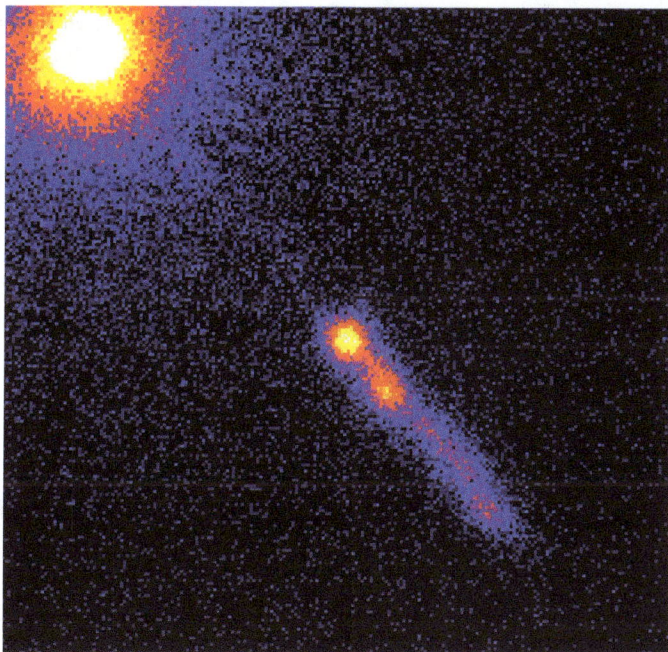

Figure 2. An X-ray image of the quasar 3C273 taken by the Chandra X-ray observatory. This quasar, which was first discovered in visible light by astronomer Martin Schmidt, constituted a breakthrough in the study of black holes. The large spot on the upper left is X-ray radiation emitted from the vicinity of the black hole. The elongated spot in the center of the image is radiation emitted from matter that was ejected by the black hole. Courtesy of NASA/CXC/SAO/H.Marshall *et al.*

about 10% of all quasars in the universe emit strong radio waves. As the American space program developed and research satellites were launched, it was found that some of the quasars also emit X-rays (see example in Figure 2), gamma rays, and infrared radiation, in addition to visible light and radio emission. Many more years passed before scientists understood that, despite observational differences among them, quasars, radio galaxies, and all of the other objects that were identified as strong radio sources were all connected to the same physical object — a giant black hole that resides at the center of a galaxy. The general name given to this diverse astronomical phenomenon is *active galactic nuclei*. As the name indicates, these objects are all related to galaxies that, unlike ordinary galaxies, exhibit irregular activity: they are brighter than usual, they exhibit rapid temporal variations, they spew matter from their centers, and more.

Up to date, nearly 200,000 quasars have been discovered, the most distant of which are located literally at the edge of the universe, about 10 billion light years away. The significance of this is that the radiation we are measuring today was emitted billions of years ago and constitutes historical record of the quasar. Quasars are also among the brightest objects in the universe; a quasar is 1,000 times brighter than an average galaxy like our Milky Way, and about 10 trillion times brighter than the Sun.

1. Quasars and the Great Gluttony

The idea that quasars are of cosmological origin and that the redshift of the spectrum results from the expansion of the universe was controversial for some time after the discovery. The strongest argument against this idea was that for such distant sources, the energy inferred from the observations cannot be supplied by means of nuclear processes like those that take place in the Sun and in other stars. It was customary at that time to believe that these nuclear processes were the most effective way of producing energy. It turned out that the quasars discovered were not only very bright, but also relatively small. Some of the quasars observed displayed brightness variations — at times they looked brighter and at other times paler. These changes occurred within very short time periods, sometimes days and even hours. Since changes that are faster than the light crossing time of source, that is, the time it takes the emitted light to travel from one end of the quasar to the other end, are not possible, the obvious conclusion was that the size of the quasar did not exceed that of our solar system. How, then, can a radiation source that small emit an amount of energy that is tens of times greater than the total energy emitted by an average galaxy? The light emitted by a typical galaxy is the accumulated light from the hundreds of thousands of stars in the galaxy. If the nuclear fusion mechanism was indeed the most efficient energy-producing mechanism in the universe, as scientists believed at the time, then no object whose mass does not exceed the mass of an average galaxy could emit more light than galaxies emit. But the fact that despite the relatively small size of the quasars discovered, the brightness measured on some of them was tens of times greater than that of an average galaxy (and especially the galaxy in which the quasar resides), required a mechanism that was more efficient than nuclear fusion.

Several alternative suggestions were made, including that the redshift of the emitted light is not related to the Doppler effect, but rather is caused by an extremely strong gravitational field, and that quasars are made of antimatter; these suggestions did not, however, gain much support. Another idea was that quasars are in fact white holes — a kind of region in space that does not enable matter or radiation to penetrate into it, but only to escape. As explained in detail in Chapter 15, the existence of white holes is theoretically possible. This option describes a mathematical solution to Einstein's field equations with properties that are the opposite of those of black holes. In fact, they are black holes in which the direction of time is reversed. But although such solutions are theoretically permitted, no natural process is known that can create white holes, and the possibility of their existence in nature is mere speculation. This suggestion was therefore also dismissed eventually.

What then is the source of a quasar's energy? The accepted explanation today is that the radiation is emitted from matter that was absorbed into a giant black hole located at the center of a galaxy. The existence of black holes of this kind was already discussed in previous chapters, and as was mentioned, their mass ranges between a million to a billion times the mass of the Sun. Because of the black hole's strong gravitational field, matter located in the outskirts of the galaxy is attracted to its center. Matter that come close to the black hole carries with it angular momentum, and so it does not immediately fall in, but rather begins swirling around the center at enormous speeds, like the vortex of water that forms in the sink or around drain holes. Black holes are, apparently, the largest drain pits in the universe, and it is through them that galactic matter pours into the region of the spacetime that is causally disconnected from the rest of the universe.

The matter swirling around the black hole creates an *accretion disk*, as it is called in the professional jargon. Clumps of matter in the disk collide with one another and the friction that results decelerates the revolving matter and causes it to spiral inward toward the center of the system. The friction also significantly raises the temperature of the matter in the disk. Due to the high temperatures, the hot matter emits strong radiation as it is advected into the center. This radiation is the quasar radiation observed at the center of the galaxy. The accreted matter eventually penetrates through the event horizon, is swallowed by the black hole, and becomes an inseparable part of it. This process, in which matter is absorbed, leads to the continuous growth of the black hole's mass, and some believe that this mechanism is responsible for the existence of black holes with such huge masses (see Figure 3).

The intensity of the radiation emitted from the disk depends on the black hole's accretion rate. The accretion rate may be calculated by measuring the brightness of the radiation emitted from a quasar located at a measured distance, provided the mass of the black hole is known. These measurements are highly important to the understanding of the disk structure, the viscous mechanism, the emission processes, and more. Measuring the accretion rate in a large sample of active galaxies will also enable us to understand what stage of cosmic evolution the black holes were created at, and what their evolution track is. This is why tremendous observational efforts were invested, and are still being invested, in performing such

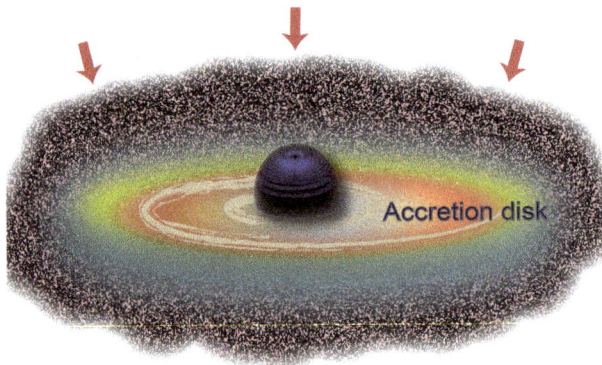

Figure 3. A schematic illustration of a black hole in the center of a galaxy accreting matter from a disk that surrounds it. The red arrows represent the flow of fresh matter from outer regions of the galaxy to the accretion disk around the black hole in the center. The inner part of the accretion disk heats up to high temperatures, and emits the quasar light that is seen from afar.

measurements on as many quasars and other kinds of active galaxies as possible. But in order to know the accretion rate of a quasar of known brightness, it is first necessary to measure the mass of the central black hole. How then can a black hole be "weighed"? One method suggested is to measure the velocity of the clouds of matter that revolve around the black hole and their distance from it, and then, using the Kepler laws, to deduce its mass. The velocity of these clouds can be deduced by measuring the spectrum they emit and applying the Doppler effect. Measuring the distance of the clouds from the black hole is a greater challenge. In a special project conducted over several years at the Tel Aviv University in Israel, headed by Professor Hagai Netzer, the mass of several black holes was determined using a method called *reverberation mapping*. This method is based on measuring the time differences between radiation that arrived on Earth directly from the disk surrounding the black hole and radiation that was emitted from the disk and was scattered in our direction by the clouds. The difference in the arrival times of the direct and scattered radiation can be readily translated into the distance between the clouds and the black hole. Researchers succeeded in estimating the accretion rate of systems in which the mass of the black hole was measured. Now the research group in Tel Aviv and other researchers around the world are working hard on applying the measurements to a larger sample of active galaxies.

2. The Spearhead — Extragalactic Jets of Matter and Radiation

So far we have focused on a description of quasar characteristics. We mentioned that some of the quasars, about one tenth of them, are bright radio sources, and we also mentioned radio sources that are not quasars, such as radio galaxies; we did not, however, elaborate on the nature of the radio emission observed in these objects.

The rapid development of radio interferometry techniques in the 1960s and 1970s, and particularly the development of the VLBI network mentioned in Chapter 6, Section 5, enabled scientists to map radio sources with unprecedented accuracy. It was found that sources that previously appeared to be point sources have complex and interesting structures.

Specifically, it was possible to discern a pair of thin jets emanating from the center in opposite directions, which end in giant lobe-like structures. The length of the jets can reach hundreds of thousands of light years. For the sake of comparison, the size of the Milky Way galaxy in which we live is close to 100,000 light years. The size of the radio lobe itself is about the size of an average galaxy. The jets in many cases are asymmetric — one is brighter than the other, and sometimes one of the jets cannot even be seen, although the lobe at its end is clearly visible (see Figure 4). A family of sources in which lobeless symmetric jets are seen also exists. Observation through an optical telescope reveals that the jets emanate from the center of a galaxy. In certain cases, a quasar is seen at the center from which the jets emanate. In fact, it turns out that a quasar is present in the center of all of these systems, but in some of the objects it is concealed from the observer's sight by a huge torus of dusty matter.

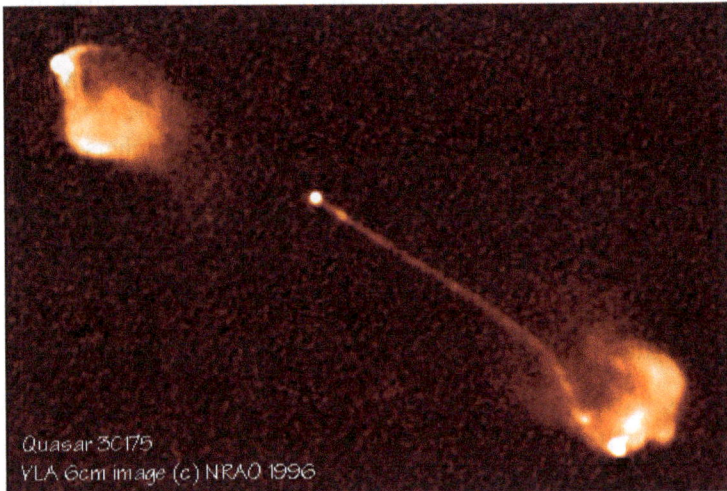

Figure 4. A radio image of the quasar 3C175. A giant black hole is located at the center of the system. The photo shows a jet of matter traveling at close to the speed of light. The counter jet is not seen in the photo due to the relativistic beaming effect. The radio lobes at the ends of the jets are powered by shock waves formed through the interaction of the supersonic jet with the ambient gas. The distance between the lobes is about three times the size of our Milky Way galaxy. Image Courtesy of NRAO/AUI.

An obvious question that arises is: What is the origin of the radio lobes? A comparison of the observations with theoretical calculations and numerical simulations indicates that the lobes result from the interaction of the jets with the ambient medium. The jets are ejected from the central black hole at velocities that approach the speed of light. Once the fast jet encounters the resistance of the external medium, it cannot continue moving forward at the same speed, and some of the matter within it spills over to the sides. Since the jet travels at speeds that are much higher than the speed of sound, a shock wave is formed at the jet's head, like a shock wave that a fighter jet creates when it exceeds the speed of sound. The matter that has spilled from the jet is heated to very high temperatures by the shock wave and is ionized. The heating process creates enormous pressure that causes the shocked matter around the jet to expand to the sides in a kind of bubble, like an inflating balloon. The jet that penetrates into the bubble plays a role similar to that of air penetrating through the mouth of a balloon when it is inflated. This inflating bubble of hot gas is the lobe. The hot gas within the bubble cools down while emitting synchrotron radiation, which is the source of the lobe's radio emission. The bubble emits a broad spectrum, and may occasionally be seen also in infrared images and in X-ray images. The atomic mushroom that appears immediately following the detonation of a nuclear bomb is a miniature example of such a system; the mushroom is actually a bubble of expanding hot gas, created by the blast wave caused by the explosion, similar to the gas bubble at the head of the jet.

In radio sources where no lobes are seen, the jets travel at subsonic velocities, that is velocities that are lower than the speed of sound. In such cases, the low jet velocity does not enable the formation of a hot supersonic bubble at the jet's edge. The gas that encounters the jet compresses a little, but remains relatively cold and therefore does not emit radiation. Why are there two families of jets, super- and subsonic? The answer to this question is still unclear. According to one version, the symmetric, lobeless jets propagate in a medium that is much denser than the medium in which the lobed jets propagate. Because of the relatively high density of the external medium, they decelerate very fast to subsonic velocities. Observational evidence exists of some of these jets

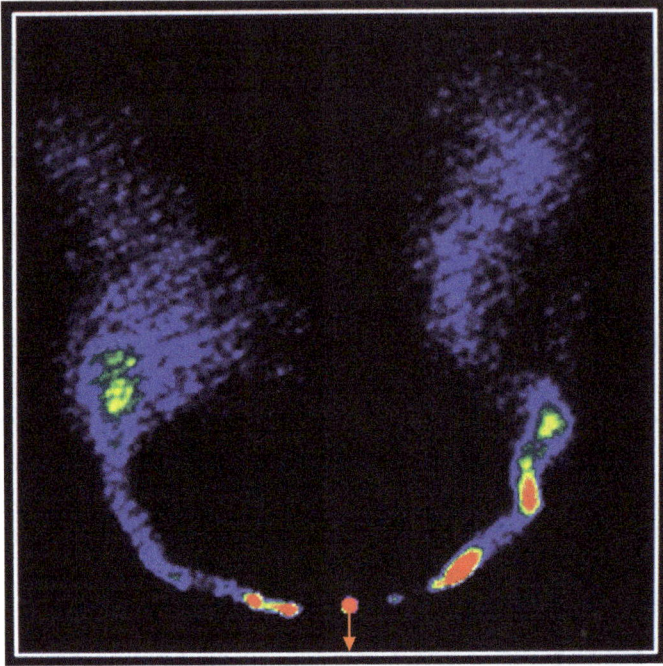

Figure 5. A radio image of a jet emanating from the center of the NGC 1265 galaxy. The galaxy itself is moving in the direction of the orange arrow at a velocity of 2,000 km/sec relative to the external medium, causing the jet to warp as seen in the photo. Courtesy of NRAO/AUI and C. O'Dea & F. Owen.

traveling very close to the black hole at velocities that are close to the speed of light, and only then do they decelerate to subsonic velocities.

Sometimes the galaxies themselves move relative to the surrounding intergalactic medium. In such a case, the jets that emanate from the center of the galaxy might warp due to friction with the matter in which they are moving, similar to the warping that happens to a taut rope or kite string when a strong wind is blowing. Figure 5 shows a good example of this: Some scientists believe that the blast waves created at the edges of supersonic jets are strong enough to accelerate protons and other atomic nuclei to extremely high energies, dozen of times higher than the energies that can be reached in the Large Hadron Collider located in Switzerland. These scientists claim that the source of ultra-high energy cosmic rays is the lobes seen in the radio images of supersonic jets. Opinions among scientists are divided in this matter. The cosmic rays may be produced in the inner regions of the jet, which emit the gamma rays observed in some of the radio sources, and it is entirely possible that the source of the ultrahigh energy cosmic rays is altogether different; possibly gamma ray bursts or magnetars, as will be described later on.

3. Faster Than Light?

A strange and extremely interesting phenomenon observed in many radio jets is *superluminal motion*, which is an apparent motion at a velocity that exceeds the speed of light. When the radio image is sharp enough, it is clear that the jets are not completely uniform, but rather exhibits a clumpy structure — very bright regions alongside darker regions (as can be seen, for instance, in Figure 6). These blobs, or "knots" as they often called, are traveling outward from the center from which the jets emanate. To measure the velocity at which the blobs are moving, scientists take a series of radio images of the same region at intervals of several months. The velocity of the blobs can be readily derived from the change in the location of a given blob and the time interval during which this change occurred. In fact, since as Earth-based observers we lack the dimension of depth, we actually can only measure the projection of the blob's motion on the plane perpendicular to our line of sight (the plane of the sky), so that the velocity measured in this method is the projected velocity.

The surprising outcome is that for many sources these features are seen to be traveling at a speed that is much greater than the speed of light. The average velocity measured in large samples of radio sources, is 10 times the speed of light (see example in Figure 6), and in several of the sources, velocities were measured that exceeded even 50 times the speed of light. How is this possible? Does this not contradict the theory of relativity, which claims that the speed of light is the highest speed in the universe?

As a matter of fact, the apparent superluminal motion is an optical illusion caused by the finite speed of light, as explained in Box 1. Even though radio knots in the jet appear to the observer to be traveling faster than light, there is no real contradiction with the theory of relativity, since information cannot be transferred through this movement. In reality, the blobs are

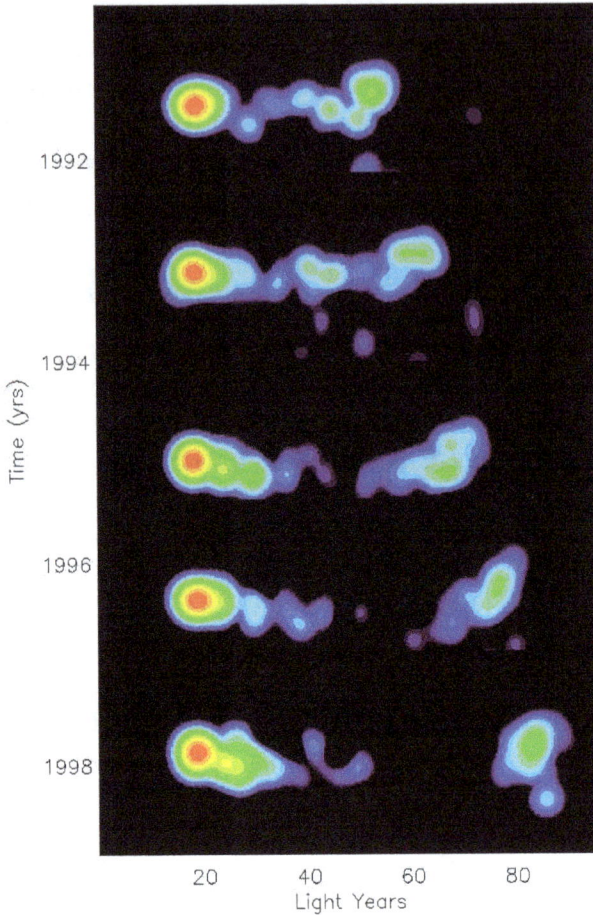

Figure 6. A series of radio images of a jet emitted from the quasar 3C279 taken between 1991 and 1998. The y-axis denotes the year each picture was taken and the x-axis denotes the distance in light years. The rightmost blob was spewed from the center of the quasar (the red circle on the left) and traveled outward. Its velocity, as can be measured from the figure, is about 10 times the speed of light. The true velocity calculated after taking the Doppler effect into account is "only" 0.997 times the speed of light. Courtesy of Glenn Piner, Whittier College. Data was taken with the NRAO's VLBA.

moving at a speed that is very close to the speed of light and to an observer they seem to be moving faster than light because of the apparent compression of time duration (an effect that is related to the Doppler effect). The apparent velocity measured depends on the true velocity of the blob and on

Box 1. Moving Radiation Sources and Time Compression

To understand the compression of time duration measured by an observer who detects emission from a radiation source that is traveling toward him, imagine the following thought experiment: a physicist is in a spaceship that is equipped with a radio transmitter and a stopwatch. The spaceship is at rest relative to an observer located on Earth, and is transmitting radio pulses exactly every 10 sec, as illustrated in the upper figure. The observer on Earth is measuring the time intervals between subsequent pulses with his stopwatch and is obtaining readings of exactly 10 sec, as expected.

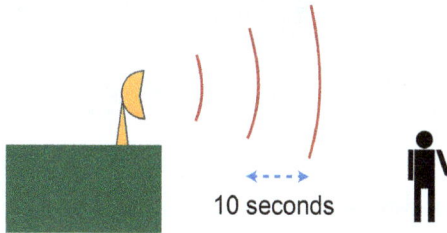

10 seconds

At this point, the physicist in the spaceship fires up the engines and accelerates the spaceship so that it is traveling towards the observer at a velocity of 0.9 the speed of light. It still continues to transmit pulses at exactly the same pace as before. The question is, what will the time intervals between pulses, as measured by the observer on Earth, be now? The answer involves an important result of the relativity theory, which is that the speed of light is universal. Specifically, electromagnetic radiation, including the radio pulses in the above experiment, always travels at the speed of light, regardless of the speed at which the radiation source is traveling. Hence, the pulses will propagate through the observer's system at the same speed whether the spaceship is at rest or in motion, and that speed is the speed of light. Since in the time between one pulse and the next, the spaceship has progressed in its journey towards the observer, the distance between any two subsequent pulses will have shortened, and as a result, so will the time interval the observer will measure, as the figure below illustrates. For the given

(Continued)

spaceship velocity of 0.9 the speed of light, the interval will shorten to one tenth of its original value. In other words, due to the spaceship's motion, the observer's clock will measure a time that is 10 times shorter than the time that actually elapses.

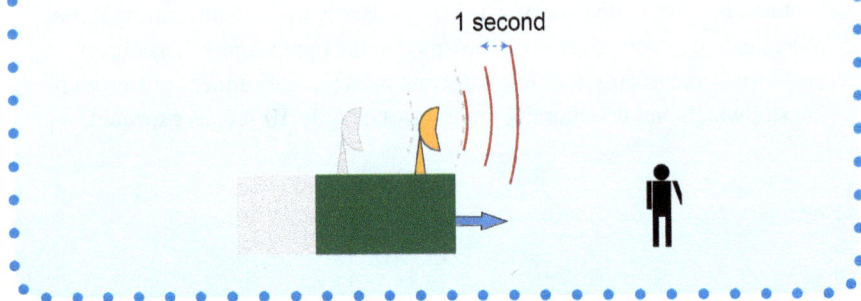

1 second

the angle between the direction of motion of the blob and the sight line. The angle at which the apparent speed is maximal depends on the ratio between the blob's velocity and the speed of light.[1] This phenomenon was in fact predicted theoretically in 1966, even before it was observed, by British astronomer and current Astronomist Royal, Sir Martin Rees.

The importance of the superluminal motion phenomenon is twofold: on the one hand it exposes, almost directly, the fact that the jets are traveling at velocities that are very close to the speed of light. On the other hand, it provides additional confirmation of the theory of special relativity. In fact, measuring the apparent velocity provides a lower limit to the true velocity of the jet. In many radio sources, the true velocity of the jet, as deduced from the measurements, seems to exceed 0.995 of the speed of light. In the past two decades, superluminal speeds have also been measured for other astronomical objects, particularly gamma ray bursts and a class of galactic sources called *microquasars*, which will be discussed later on.

Superluminal motion is also essentially related to the asymmetry of oppositely oriented relativistic jets mentioned above. The fact that one of the jets in most of these systems seems brighter, while the other looks

---

[1] To be precise, the effect will be maximal when the ratio between the clump velocity and the speed of light equals the cosine of this angle.

faint or cannot be seen at all, stems from a relativistic effect called *Doppler beaming*. Due to light aberration, radiation emitted by a relativistically moving source is highly beamed along the source's direction of motion, as illustrated below, even if the emission in the rest frame of the source is isotropic (uniform in all directions). As a result, a radiation source that is moving toward the observer will appear brighter then if it were at rest, while if it is moving away from the observer it will appear much fainter, since the emission is beamed away from the observer.[2] This is the origin of the brightness asymmetry seen in the radio images — the jet moving towards us seems especially bright, while the jet moving away from us looks especially faint (see Figure 7). The Doppler effect that

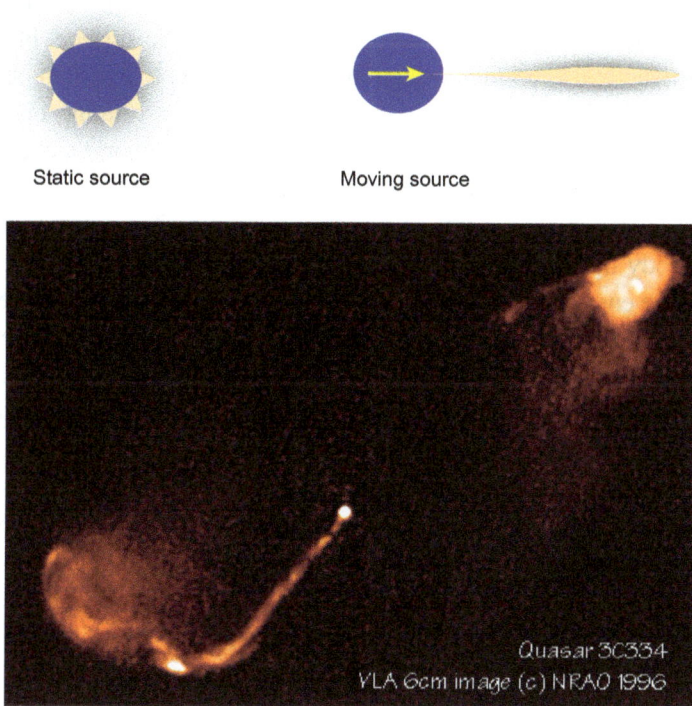

Static source   Moving source

Figure 7. The beaming effect: The jet on the left is moving towards us, and therefore appears bright. Its counterpart on the right is moving away from us, and so looks much paler. The exposure time in this photo is too short to enable us to discern the jet that is moving away. Image Courtesy of NRAO/AUI.

---

[2] In addition to light aberration, the rate at which photons are emitted is affected by time dilation and their energy is Doppler shifted, which enhances the effect.

causes the superluminal motion and the beaming effect are interrelated and, in fact, constitute two aspects of the same phenomenon.

4. Magnetic Pumps of Rotational Energy

How are the jets formed? What accelerates them to velocities that approach the speed of light? Scientists have been concerned with these questions since radio jets were first observed, almost five decades ago. According to the currently prevailing theory, two factors play a major role in the creation and acceleration of the jets: rapid rotation of the black hole around its axis, and magnetic fields. As we explained in Chapter 15, a rotating black hole has two special surfaces that encompass a region called an ergosphere. The inner one is the event horizon, while the outer one is the static surface. Inside the ergosphere, no body can remain stationary with respect to a distant observer. This is because the twisting of spacetime induced by a rotating black hole forces bodies in that spacetime to rotate. No force in the world can resist the rotational coercion that a black hole induces within its ergosphere. Furthermore, there are, within the ergosphere, negative energy orbits. A physical body that falls into a black hole through such a trajectory imparts to it negative energy or, in other words, causes the black hole to lose energy. The missing energy must obviously somehow reach the space outside of the ergosphere, far from the black hole, a process that requires some kind of mediator. When such an entity is present, it is in fact possible to "pump out" the rotational energy of the black hole with very high efficiency.

The mediating factor is magnetic fields! Where do these fields originate? A great deal of evidence indicates that in essentially all of the systems in which jets are seen, the central black hole accretes matter from a surrounding disk, just like in the case of the quasars discussed earlier. The rotation of the disk around the black hole creates an electric current, which in turn generates a magnetic field in the entire space. Some of the magnetic field lines penetrate the black hole through the event horizon. The magnetic field lines may be likened to hairs that emanate from the disk and the event horizon.

The twisting of spacetime induced by the black hole causes the field lines that penetrate into the ergosphere to twist as well. A small amount of the ionized matter in the disk and its surroundings slides along the magnetic field lines into the ergosphere, and from there into the black hole. Some of the electric charges that slide through the magnetic field lines into the ergosphere fall into the black hole by way of negative energy trajectories and decelerate its rotation. The rotational energy that the black hole loses flows out of the ergosphere along the twisted magnetic field lines (even through the mater itself falls inward), eventually becoming a jet of magnetized matter that emits the radiation we see in the radio images (Figure 8). The jets in this model are in fact an apparatus for pumping out the rotational energy of black holes. The twisting magnetic field lines may be likened to a giant Archimedes screw, and the black hole to a flywheel that is attached to the screw. Turning the flywheel rotates the screw, whose motion causes water to flow up from the bottom of the well.

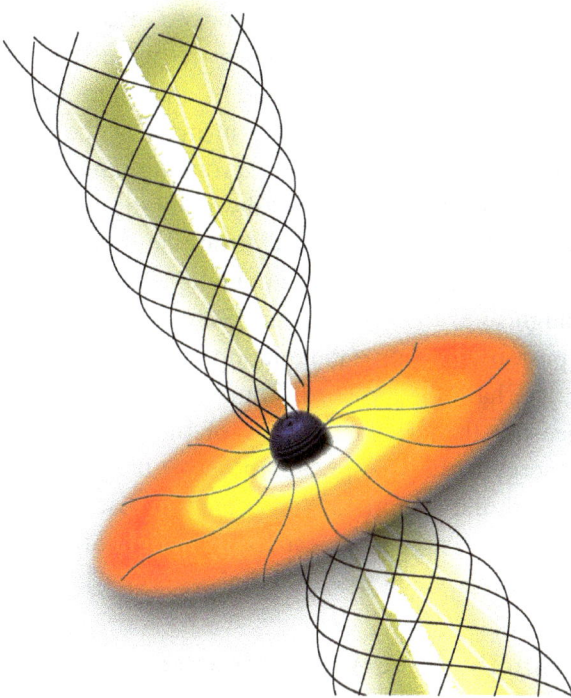

Figure 8. A schematic representation of a magnetized jet. The network of black lines represents the magnetic field lines in the jet. The rotational energy of the black hole that powers the jet flow outward from the black hole along these lines.

The rotational energy of the flywheel is converted into the energy required in order to remove the water and, if the wheel is not connected to a motor, its rotation rate will decrease over time. The water flowing up the screw is analogous to the jet that is ejected from the black hole.

Despite understanding the main points of the magnetic pumping process described above, many important details are not yet sufficiently clear. Significant progress in understanding was made in recent years thanks to the rapid development of computational capabilities and the invention of sophisticated methods that enable numerical simulations of the processes involved in the creation and acceleration of the jets. Additional development requires a comparison of the mathematical models and computer simulations with observations of jets in the vicinity of black holes. Existing technology enables to resolve only those parts of the jet that are relatively distant from the black hole.

Another important question is, why do only a relatively small proportion of quasars exhibit relativistic jets when we claimed earlier that they all harbor a gigantic black hole surrounded by an accretion disk? The current understanding is that the black hole in most of the quasars rotates slowly or not at all. If the black hole is not rotating rapidly about its axis, an ergosphere will not be created, and the pumping mechanism cannot function, even in the presence of a strong enough magnetic field. Apparently, only 10% of the quasar population have a black hole that rotates fast enough so as to create the jets. Some of the quasars may have had jets in the past, but over time the rotation of the black hole has slowed down, and the jets have disappeared. Some researchers claim, based on attempts to measure the angular velocity of black holes, that evidence exists that the energy in the jet is greater the faster the rotation of the black hole, in accord with the jet formation mechanism discussed above; however, these claims are controversial, especially in light of the methods used to measure the rotation rate of the black holes.

5. A Glance into the Innards of a Monster

If it were possible to observe the matter around a black hole directly, it might be possible to see how a jet was created and how it is accelerated. By comparing that with the theory, it would be possible to learn a great deal about the jet formation mechanism, and about the connection between the jet and the accretion disk, and maybe gain additional information about the properties of black holes. Due to the great distance of these objects from Earth, however, it is impossible to resolve such small regions, even using the best radio telescopes available. The attempt to obtain a sharp image of matter that is located near a giant black hole in a galaxy that is a billion light years from Earth (the average distance of such objects) is equivalent to an attempt to see an ant from a distance of 100,000 km (about quarter of the distance from Earth to the moon). To make this feasible, the resolution of existing telescopes must be improved at least 1,000-fold.

Nevertheless, one galaxy, named Messier 87 or M87 for short, is very close to Earth — only about 60 million light years away. At the center of M87 is a giant black hole, one of the largest ever discovered in the universe, with a mass of about six billion times the mass of the Sun. The jet that emanates from M87 is seen to extend over great distances in radio images, in visible light and X-ray images as well. The proximity of this system enables, for the first time, to glimpse into the innards of the monster. It is still not possible to really see the black hole, but it is possible to see much smaller details than are visible in the other systems. A special project conducted recently used the largest radio interferometer in the world, the VLBI, to take a series of radio images of the black hole region. In the sharpest of the images, like the one in Figure 9, areas 50 times the size of the event horizon may be discerned. The series of images was made into a movie that may be viewed on the Internet. The movie shows how matter is thrown from the black hole, accelerates, and while in motion collimates into a jet.[3] Comparing these data to models and computer simulations enables a better understanding than ever before of the jet formation mechanisms, and serves as an active research subject. The next generation of telescope, as many hope, will enable scientists to image the activity near the event horizon itself.

6. The Black Hole Silhouette

Up until today, scientists have succeeded in measuring the size of white dwarfs, and indirectly also of neutron stars. They have also managed to

---

[3] A link to the movie: http://www.aoc.nrao.edu/~cwalker/M87/index.html.

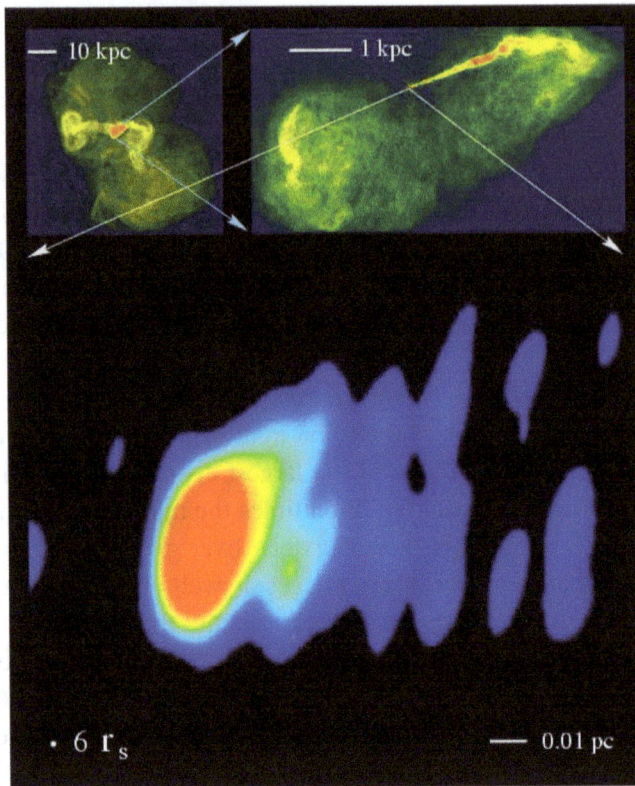

Figure 9.   The bottom figure presents a radio image of the inner most parts of a jet emanated from the center of the M87 galaxy. The black hole is located at the center of the orange spot. The radius of this spot is about 50 times the radius of the black hole. Image Courtesy of NRAO/AUI.

learn a lot about the structure and composition of these objects and to provide important confirmation of the theories that concern them. Yet nobody has yet succeeded in measuring the size of a black hole. Such a measurement, in addition to the direct measurement of gravitational waves, would be an immensely important test of the theory of general relativity.

It would seem, however, that such a measurement is impossible, since, by definition, a black hole emits no radiation and so cannot be seen. Nevertheless, nothing prevents us from seeing its shadow on a radiation source located behind it, and deducing its size directly by measuring the shadow.

It is still impossible to measure the shadow, since such an operation would require a telescope with a resolution that yet unavailable. Telescopes in the microwave range that could enable such measurements are expected to become available in the near future. Special infrared observation techniques, which so far were used to study the center of our galaxy, may also enable such measurements, after some improvements. The next generation of telescopes will, most probably, be capable of measuring the shadows of the closest giant black holes — those in the center of the Milky Way and the M87 galaxies.[4]

Measuring the shadow of a black hole is insufficient, however, if we wish to study the properties of the spacetime induced in its vicinity. Data analysis also requires a theoretical computation of the propagation of light rays near the black hole. The warping of spacetime in the close vicinity of the black hole is strong and causes significant deflection of the rays emitted from any radiation source that pass close to the event horizon, similar to the way light is refracted by an optical lens. This effect, which is called *gravitational lensing*, leads to strong distortions of the radiation source. In order to analyze observations of radiation sources that are close to a black hole, these distortions must be taken into account. Computer simulations enable scientists to create artificial (synthetic) images of known radiation sources and to use them in the future to interpret the observations. An example of such a synthetic image produced by a group of researchers at Harvard University, headed by Prof. Avi Loeb, is shown in Figure 10.

7. Particle Accelerators and Gamma Ray Jets

After the Compton Gamma Ray Observatory was launched into space in the early 1990s, it became clear that many of the then-known radio jets emit very strong gamma radiation. In fact, it was found that the jets emit most of their energy in the form of gamma rays, and only a smaller fraction is emitted as radio waves, infrared radiation, and visible light. Later, with the operation of Imaging Atmospheric Cherenkov Telescopes (IACT) and the launching of the Fermi Gamma Ray Space Telescope

---

[4]These two particular systems harbor black holes with the largest angular sizes, as measured on Earth, among all known systems; that is why most efforts are currently focused on them.

Figure 10. A simulation that depicts the shadow of the black hole located at the center of the Milky Way galaxy against the background of radiation emitted by the matter that surrounds it. Courtesy of A.E Broderick & A. Loeb.

(see description in Chapter 6), many active galaxies were discovered that were not previously known, and which exhibited strong gamma ray jets. It then became clear that the gamma ray spectrum extends to much higher energies than those measured by the Compton Gamma Ray Observatory. Up until today, over 1,000 active galaxies have been identified as bright gamma ray sources. Theoretical calculations have shown that the particles (probably electrons) emitting the gamma rays are accelerated to very high energies, tens of times higher than the energies that can be achieved in the largest man-made particle accelerators. It seems that the greatest particle accelerators in the universe were already created by nature billions of years ago.

How are the particles in the jet accelerated to such high energies? When water is pumped through a hose at high enough pressures, the hose is noticeably unstable: this is particularly evident in large fire hoses mounted on fire engines. Similarly, the jets that emanate from black holes are unstable due to their rapid flowing, as well as due to intermittencies of the injection process; they distort, and at times the flow stops altogether for short periods of time and is then immediately renewed, like water flowing from a faucet after a water stoppage. As a result of the intermittent flow, blobs of matter in the jet collide with one another, and since the jet material is traveling at a supersonic velocity, these internal collisions create strong shock waves that travel along the jet and heat the surrounding gas to very high temperatures. Most of the gas particles continue with the flow, and only a relatively small proportion of them trapped by the shock wave, specifically by the magnetic fields that are swept along with the jet, and they collide time and time again with the matter on either side of the shock front. Each collision accelerates the trapped particle and adds to its energy, and so, after several collisions, the particle's energy grows to a much higher value than its initial energy — sometimes even a billion times and more higher. The more collisions, the greater the energy. This is a bit like a game of Ping-Pong: each time the ball hits one of the player's bat, it is accelerated and changes direction (see Figure 11).[5]

The energy a trapped particle gains originates in the jet; as a result of the repeated collisions of the accelerated particles, the jet loses kinetic energy and decelerates down. Although only a small proportion of particles are trapped in the shock wave and are accelerated, they carry with them a significant fraction of the overall energy of the jet. This may be likened to a spaceship traveling rapidly through a cloud of tiny meteorites (the size of a grain of sand). Any meteorite that collides with the spaceship is thrust forward at a higher speed than it had before the collision, and thus gains kinetic energy. The energy that the meteorite gained is lost by the spaceship. The collision of a single meteorite will not significantly affect the

---

[5]The reason the velocity of the ball does not increase endlessly is air resistance. If we could play ping-pong in outer space with perfect bats, we would discover that the ball moves faster and faster with every stroke of the bat.

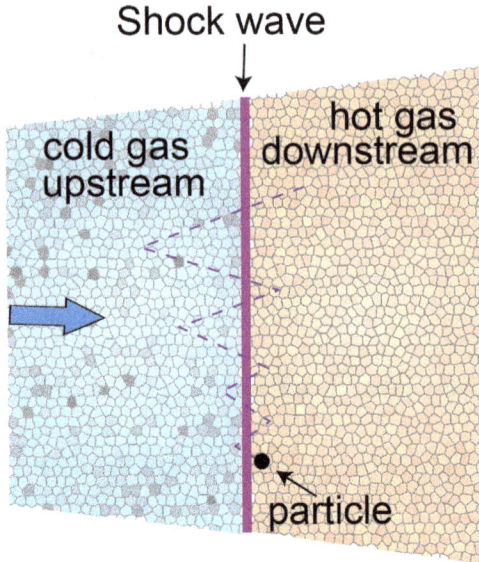

Figure 11. The dashed line represents the trajectory of a trapped particle that repeatedly collides with matter on both sides of the shock wave. The particle is accelerated with each collision, increasing its energy.

spaceship's motion, since it is tens of times heavier, but a relatively high number of such collisions will cause the spaceship to experience an accumulative loss of energy and a considerable deceleration.

What is the maximum amount of energy that a particle can gain in this manner? The answer depends on the kind of particle and on the conditions that prevail within the jet. For electrons, this energy is limited by radiation losses. An accelerated electron loses part of the energy it gained, through the radiation it emits. This is of course the energy that we can see. The greater the electron's energy, the higher the rate of radiation loss. When this rate equals the rate at which the electron gains energy as a result of its collisions with the shock wave, the electron stops accelerating. In this state, the additional energy the electron gains in each collision is immediately emitted as radiation. For heavier particles, protons in particular, the critical energy at which the radiation loss rate equals the acceleration rate is much higher than for electrons. For such particles, a different limit may exist. The greater the particle's energy, the harder it is to keep it trapped.

At some point, when the energy the particle acquires exceeds a certain value, it will simply escape from the jet to the space outside the jet, and the acceleration process will come to a halt. The energy at which this happens depends on the size of the shock wave and on the strength of the jet's magnetic fields (by which the particles are trapped, as mentioned before).

For typical values of the shock wave dimensions and the magnetic field intensity measured in some of the jets, the maximum energy a proton can acquire in the process described above matches the energies measured for the ultra-high energy cosmic rays that penetrate the galaxy from the outer space. This is why some researchers tend to believe that the internal shock waves in the giant jets observed in active galaxies are the source of these ultra-energetic particles. Other researchers reject this notion and claim that the number of active galaxies that are close enough to the Milky Way is too small to account for the measured flux of these cosmic rays. It does not look like this debate will be decided any time soon. The acceleration mechanism described above is not unique to jets, but rather works in every system in which strong shock waves are formed. An important example, which will be discussed later on, is shock waves that create supernovae, where cosmic rays are apparently accelerated to medium energies (that are much lower than those of the ultra-high-energy cosmic rays).

8. The Story of Microquasars

In 1994, a double radio jet whose properties were similar to those seen in quasars and active galactic nuclei was discovered. Superluminal motion of radio knots that were thrust from the center of the system was also measured, as described above. However, data analysis revealed with absolute certainty that the jet is not extra-galactic, like those that were known at the time, but rather is located within our galaxy, at a distance of 30,000 light years from Earth. The immediate conclusion was that the size of the jet does not exceed one hundredth of a light year, in other words, less than a millionth the size of a typical extra-galactic jet. It was also found that the center from which the jets emanates coincides with a binary star system that belongs to a class of objects called *X-ray binaries*, whose characterizing property is the strong X-ray emission. X-ray binaries are made up of a pair of stars — an ordinary star and a black hole or neutron star — that revolve around each other in a kind of celestial dance. Due to the intense attraction between the stars, matter is torn from the ordinary star and falls onto the compact star. If the compact star is a black hole, the matter accumulates around it and creates an accretion disk. The inner regions of the disk heat up as a result of the friction and emit the X-ray radiation that we see, like in the case of quasars. When the counterpart is a neutron star, the matter accumulates on the edge of the neutron star, is compressed and becomes very hot, emitting X-ray radiation as a result. In this case too, a disk is formed in some of the cases. The essential difference between accretion to a black hole and accretion to a neutron star is that in the former case the overflowing matter is swallowed beyond the events horizon, and in the latter case the matter eventually collides with the neutron star and accumulates on its surface. This leads to differences in the observational characteristics of the two kinds of systems, particularly in the X-ray emission characteristics.

An analysis of the X-ray emission characteristics of the binary star system identified at the center of the radio jet discovered in 1994 led to the conclusion that the ordinary star revolves around a black hole whose mass exceeds 10 times the mass of the Sun, and it was conjectured that the jet probably emanated from the black hole vicinity. Due to its great similarity to the jets observed in quasars, this system was called a *microquasar* (see Figure 12). The prefix *micro* was intended to indicate the system's size scale (about one millionth the size of a typical quasar). Since then, additional microquasars have been discovered in our galaxy, and several dozens have been documented up to date. It was also found that the jets are launched only when the X-ray luminosity of the system greatly diminishes, indicating a significant decline of the mass accretion rate. When the X-ray luminosity is especially high — evidence of a high accretion rate — the jet injection process stops for a reason that is not yet entirely understood.

Where do these systems originate? We mentioned earlier that most of the stars in the universe are in binary systems, in which two ordinary stars revolve around each other. When the time comes for the heavier companion to die, it collapses and becomes a black hole or a neutron star. If the explosion caused by the collapse of the star was not strong enough to break up the binary system, then a new binary system will form that consists of an ordinary star and a compact object — a black hole or a neutron star. In the system discovered in 1994, the collapsed star became a black hole.

Figure 12. Illustration of an X-ray binary system: A black hole accretes matter from an ordinary star. The matter accumulates in the accretion disk, and from there it slides toward the black hole, emitting strong X-ray radiation. Jets of matter are sometimes expelled from the black hole (not shown here), similar to what happens in radio galaxies and quasars. Illustration: NASA/CXC/M. Weiss.

In time, the black hole's companion will also die and become a neutron star, and so a binary system will probably ensue in which a neutron star revolves around a black hole. This, however, will occur only in another several million years.

In some of the X-ray binaries in which the compact object is a neutron star, the matter that accumulates at the edge of the neutron star heats up to a high enough temperature so as to cause nuclear reactions on the stellar surface, like those that take place in novae, but much more intense. This process leads to a huge release of X-ray emission over very short durations that typically last only a few seconds. In some of the systems, this process repeats itself, sometimes at intervals of several hours. These objects, which were first discovered in the 1970s, are called *X-ray bursts*.

Chapter 17

Death and Birth — The Tale
of Cosmic Explosions

1. Was a Star Born?

One dark evening, while watching the sky on his way home, Tycho Brahe, one of the more prominent astronomers of the 16th century, noticed the appearance of a new, especially bright star in the constellation Cassiopeia. Tycho was greatly surprised since, at that time, the stars were believed to be fixed, eternal and unchanging luminaries. At first, Tycho thought it was a different celestial body or an atmospheric disturbance, which appeared as part of the constellation Cassiopeia merely by chance. However, over the many observations he conducted after the appearance of the vision until it faded away about 18 months later, it became clear that the new star did not change its position relative to the other stars in the constellation, as would be expected if it were a planet, a comet, or some other nearby object. Tycho concluded that the star that appeared suddenly and disappeared gradually was part of the group of fixed stars. In 1573, Tycho published this discovery in his book *De Nova Stella* (*The New Star*).

Today we know that the star that was "born", which Tycho discovered, was not really a star but rather an explosion, probably of a white dwarf, called type Ia supernova. The remains of the explosion Tycho saw can still be seen today, almost 450 years after it occurred (see Figure 4). In 1604, Johannes Kepler discovered another supernova. This supernova also left remnants that can be seen today. No additional supernovae have been discovered in the Milky Way galaxy since then, but that does not mean that explosions of this kind have not occurred. Some of the supernovae apparently occur in star forming regions (also called stellar nurseries) in

Figure 1. The yellow arrow points to a supernova discovered in 1994 in the NGC4526 galaxy by the Hubble Space Telescope. At the peak of its explosion, the supernova was brighter than the entire galaxy. Courtesy of NASA, ESA, the Hubble Key Project Team and the High-Z Supernova Search Team.

which the density of matter and dust are sufficiently high so that they absorb most of the light emitted by the supernova. According to various estimates, supernovae occur in our galaxy about once every 30 years. Such explosions are also observed throughout the universe at a very high frequency. Explosions that take place in galaxies that are not too far away from Earth can be detected even with a small amateur telescope, and many supernovae have indeed been discovered by amateur astronomers around the world.

2. Cosmic Explosions

Our daily experience teaches us that the detonation of a bomb is accompanied by a tremendous flash of light followed by a deafening noise. We experience something similar during thunderstorms as well;

first a burst of light — lightening, followed by a rumbling noise — thunder. What causes this phenomenon?

In the course of the explosion, prodigious amounts of energy are released within a short time and in a relatively small volume. In the case of a conventional bomb, the energy is stored in the explosives in chemical form. Detonation of the bomb causes very rapid combustion of the explosives and the heating of an air bubble in the explosion area. Heating of the bubble excites the air molecules and causes the emission of radiation which we see as a flash of light. The great pressure created in the bubble following the rapid heating causes it to inflate at a supersonic speed, which results in the creation of a blast wave that propagates outward from the explosion area. Most of the destruction that accompanies explosions is caused when the blast wave encounters structures, vehicles, human being, and so on. A similar process takes place when an atomic bomb detonates, but here the explosion energy is nuclear energy, which is obtained from the fission of a uranium atom. In the case of lightening and thunder, the air is heated by high-energy particles that are accelerated in the electric field created in the atmosphere during the thunderstorm. Lightening is the light that the heated air molecules emit, while thunder is the sound wave. In all of these cases, the flash of light is seen before the sound of the explosion is heard, since the speed of light is greater than the propagation speed of the blast wave (which is approximately equal to the speed of sound in air). During its propagation, the blast wave weakens, until it dissipates. The impact radius of an explosive is the distance from the explosion beyond which the blast wave is too weak to cause any damage. This radius is larger the higher the energy of the explosion. The impact radius of an atomic bomb is so much greater than that of a conventional bomb because the amount of energy stored in nuclear fuel is tens of times greater than that which can be stored in chemical form in the same amount of matter.

A supernova is a tremendous cosmic explosion. Here too, a huge amount of energy is released within a short period of time in a small volume. As mentioned earlier, there are two types of cosmic explosions, or supernovae, in the universe: type II supernovae, which are related to the death of an ordinary star, and the birth of a neutron star or a black hole, and type Ia supernovae, which are most likely related to an explosion of a white dwarf. In the first instance, the source of energy is the gravitational potential energy of the collapsed core. This kind of explosions comes with an important warning sign, which will be discussed later on. In the second instance, that of a white dwarf explosion, the energy source is nuclear, like a cosmic atomic bomb. In both cases, the burst of light we see is emitted from galactic matter that was heated to high temperatures as a result of the explosion. The blast wave created by the explosion continues to propagate through the galaxy for hundreds of thousands of years, and it may be observed.

3. A Neutrino Signal from a Collapsing Core

The moment the pressure in the center of the star drops, after the nuclear reactor shuts off, the core of the star, which contains about one tenth of the mass of the original star, begins collapsing. The core material falls into the center at increasing velocity due to the mutual gravitational force that acts between them, and the gravitational energy of the core becomes the kinetic energy of the matter that is accelerating inward. The collapse of the core occurs much quicker than the shrinking of the stellar envelope, which contains the rest of the matter, leading to the detachment of the star's core from its external envelope. During the collapse, the core heats up as a result of random collisions between the clumps of core material falling into the center. As a result of the great heat created, the core's atomic nuclei decompose into their various components — protons, neutrons, and electrons — and prodigious radiation is emitted. The high density of the matter in the core prevents the radiation from escaping, and it remains trapped. In addition, because of the enormous pressure that prevails in the core, protons and electrons merge and form additional neutrons, and neutron-rich matter is created. In this process, which is the inverse process to beta decay (see Chapter 7), anti-neutrinos (that is, antiparticles of neutrinos) are emitted, which, unlike the electromagnetic radiation, can escape freely. During the collapse, the density of the collapsing matter increases rapidly, until finally it reaches a value at which the quantum pressure of the neutrons is high enough to balance the gravitational force, and the collapse of the core stops all at once. The entire collapse process lasts less than a second. If the mass of the core exceeds

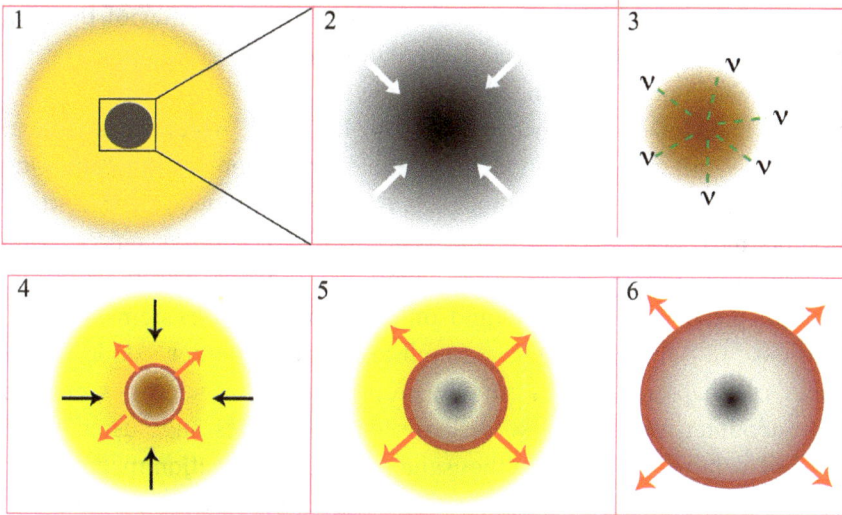

Figure 2. Stages in the collapse of a star: 1. The iron core (gray circle) immediately before its collapse and detachment from the yellow envelope of the star. 2. The collapse of the core and heating of the core material. 3. End of the collapse and the creation of a hot neutron star, which cools down while emitting a burst of anti-neutrinos. 4. A shock wave is created due to the heating of the inner part of the stellar envelope. 5. The shock wave propagates outward through the envelope. 6. The shock wave breaks out of the envelope into the interstellar medium.

the maximal mass possible for a neutron star, the collapse will continue until a black hole is formed. Figure 2 explains various stages during the collapse of a star.

In the initial seconds following the collapse, a lump of very hot neutron matter is created (proto-neutron star), which later cools and becomes a neutron star. The temperature of the neutron gas can reach 100 billion degrees, 10,000 times the temperature in the center of the Sun. The hot lump of neutrons emits anti-neutrinos and, as a result, gradually cools down. The initial cooling lasts a couple of seconds, during which most of the gravitational energy stored in the core prior to the collapse is released in the form of anti-neutrinos. Some of the energy emitted, about 1% is absorbed by the inner part of the stellar envelope, which continues to shrink slowly. This causes rapid heating of the envelope and the creation of a shock wave, which propagates outward from the center of the star, making its way through the relatively dense envelope of the star, until it finally breaks out into the medium outside the star.

The burst of anti-neutrinos from the hot core, which lasts several seconds, is the most important theoretical prediction — the "smoking gun" of the stellar evolution model. It is therefore no wonder that many attempts have been made over the years to discover these anti-neutrino particles; these attempts, however, have all been futile. The greatest difficulty in discovering neutrino (and anti-neutrino) particles stems from the great ease in which they pass through any kind of matter. As described earlier in Chapter 7, only a minute proportion of the neutrinos emitted ever encounter matter, and so their discovery requires huge detectors in terms of volume, that contain thousands of tons of detection matter. Such experiments are usually located within deep mines so as to prevent possible disturbances caused by cosmic rays.

A breakthrough, which constitutes an important landmark in the long history of research into the structure and evolution of stars, took place in 1987 (see Figure 3). A star, located in a nearby galaxy by the name of the Large Magellanic Cloud, collapsed to its death and exploded. The supernova was discovered simultaneously in several locations around the world, less than 24 hours after the light emitted from it began to reach Earth. Two amateur astronomers photographed the region where the explosion took place on 23 February, within an hour of each other. In the first photo, taken at 9:30 UTC (Coordinated Universal Time), the supernova was not yet visible while in the second photo, taken one hour later, it was apparent. Hence, the first light reached Earth between 9:30 and 10:30 UTC. The explosion itself occurred about 167,650 years earlier — this is the time it took the light to traverse the distance from the Large Magellanic Cloud to Earth.

This event took place close enough to Earth so that it was possible to discover the anti-neutrino particles using existing means. Indeed, careful examination of the data revealed that in a burst that lasted 13 seconds, about 11 anti-neutrinos were detected by the Kamiokande detector installed in the Kamioka mine in Japan (a photo of the detector was presented in Chapter 7). A similar reading was obtained by two other neutrino detectors, one in the United States and the other in Russia. The

Figure 3. Supernova 1987A — Before and after: The bright spot of light in the center of the left photo is the supernova observed in February 1987. The image on the right presents the same region before the explosion. The white arrow indicates the star that collapsed and created the supernova. Courtesy of Australian Astronomical Observatory/David Malin Images.

quantity of particles detected enabled scientists to calculate the total amount of energy emitted in the explosion. The value obtained was in good agreement with the theoretical prediction and served as final confirmation of the stellar evolution model. The arrival times of the anti-neutrinos recorded by the detectors indicated that the anti-neutrinos reached Earth about three hours before the light from the supernova did, implying that there is a lag between the time of the explosion and the emission of the light. The reason for this delay is that the light was not emitted immediately upon explosion, but only after the shock wave broke out of the stellar envelope.

According to the theoretical prediction, a neutron star should have been created in the center of the supernova discovered in 1987. But researches conducted using the space telescope and other means failed to reveal any evidence of the expected neutron star. Several hypotheses were raised regarding this deficiency. According to one hypothesis, the star is shrouded by very dense clouds of dust and gas that absorb its radiation and make it difficult to detect. According to another conjecture, during the explosion, matter from the stellar envelope fell onto the neutron star and caused it to collapse into a black hole. The third possibility proposed was that the explosion created a quark star.

4. Shock Breakout and Light Emission

As mentioned earlier, immediately following the explosion, the shock wave that is created starts propagating through the dense envelope of the star. The shock wave heats the gas in the envelope to high temperatures, and the hot gas emits intense radiation. Due to the large density of the envelope, however, the radiation remains trapped and cannot escape. Only several hours (to days, depending on the type of progenitor) after the explosion, does the shock wave break its way out of the dense stellar envelope, and begins propagating through the interstellar medium. The first light of the supernova is emitted upon emergence of the shock wave at the edge of the stellar envelope. However, since the density of the emitting matter is still high there, most of the radiation remains trapped in the expanding shocked shell and only a relatively small amount is emitted.

After breaking through the stellar envelope, the shock wave continues on its journey through the external medium. Since the density of the expanding shell decreases with the distance from the core, the matter becomes increasingly "transparent", and a greater amount of light is emitted the longer the shock wave continues to propagate. A person standing on Earth and observing the explosion area will see a light source that becomes brighter over time, like a star being born, just like the Chinese astronomers who observed the explosion that begot the Crab Nebula saw, and as Tycho Brahe saw in 1572. At some point, the source will achieve its peak brightness, when the matter becomes completely transparent to the emitted light. The light will then diminish slowly due to the weakening of the shock wave, which continues to propagate, and the radioactive decay of iron peak elements produced in the explosion, as explained next. Peak brightness comes several days after the explosion. The decaying process lasts a few weeks to a few months, depending on the kind of star that exploded and on the external conditions.

The high temperatures and density that prevail in the core of a collapsing star during the explosion lead to the creation of elements that are heavier than iron, such as nickel. These elements, which are expelled from the star and propagate outward together with the shock wave, are unstable and undergo radioactive decay during the propagation of the shocked shell. This radioactive decay releases a large amount of energy, which contributes to the heating of the matter. In the initial days following the explosion, the main source of energy is the shock wave itself. Later on, however, most of the energy emitted from the propagating shock wave comes from the decay of the radioactive material, and primarily from nickel. This is the reason for the relatively slow decline in the intensity of the light from the explosion. An analysis of the light curve, that is the change in brightness over time, enables to deduce the composition of the radioactive substances formed during the explosion.

5. The Luminescent Demise of White Dwarfs as Cosmic Probes

In Chapter 10, we described a binary star system in which one of the two stars is an ordinary star and the other is a white dwarf. We described how accretion of matter from the ordinary star by the white dwarf might cause nuclear explosions on the surface of the dwarf, which, to an observer on Earth, look like novae. When the distance between the stars is small enough, the accretion rate increases way beyond the rate exhibited in novae. The mass of the white dwarf increases due to the accretion, and eventually exceeds the Chandrasekhar mass. At this stage, the electron pressure cannot overcome the gravitational force, and the dwarf begins to collapse, leading to very rapid heating of the surface of the star to very high temperatures. When the temperature reaches a certain critical value, the carbon ignites and a nuclear reaction begins that causes an accelerated release of energy and, ultimately, the entire star explodes (like a huge nuclear bomb). Such an explosion, which signals the death of the white dwarf, is called type Ia supernova. Type Ia supernovae are distinct from type II supernovae, which result from the core collapse of a massive star. From an observational perspective, the primary difference between the two types of supernovae is in the characteristic spectrum: spectra of type II supernovae contain hydrogen lines while spectra of type Ia supernovae exhibit no trace of hydrogen. This difference fits well with the theoretical prediction; in the case of a type II supernova, the explosion blows off the envelope of the collapsing star, which is composed mainly of hydrogen. The hydrogen, which is heated by the shock wave, then emits the lines seen in the spectrum. White dwarfs, on the other hand, contain no hydrogen at all, but rather mostly carbon and oxygen, and so there is no hydrogen in an explosion of a white dwarf which could emit those lines. Many details related to the formation of type Ia supernovae are still unclear. The scenario suggested above is accepted by many scientists, but alternative theories also exist that include the possibility that such an explosion occurs following the collision of two white dwarfs that revolve around each other. Like in the case of a collapsing star, the explosion in this scenario also creates a shock wave that propagates through the interstellar medium for a long time after the explosion and emits radiation.

Type Ia supernovae, that are created from the explosion of white dwarfs, are of great importance in the study of the universe. As explained in Chapter 1, in order to learn about the structure of the universe and deduce its age, we must measure the distance to objects that are as distant from us as possible. Measuring the distance to faraway galaxies is one of the main challenges in cosmology, and immense resources have been invested over the years to overcome it. The accepted method is to use a class of objects of known luminosity, called *standard candles*. Although astronomers around the world have been searching for standard candles for many years, it was recently discovered that type Ia supernovae are a group of standard candles of especially high brightness that can be detected even at great distances from Earth. Even though the mechanism responsible for the white dwarf explosion is not sufficiently clear, it turned out, based on extensive observations, that the same amount of energy is always released in every explosion. This discovery has led different research groups from around the world on an intensive quest to find type Ia supernovae throughout the universe. As already mentioned, the identification of supernovae at great distances from Earth is what led recently to the conclusion that the expansion of the universe is accelerating and that its content is largely dominated by dark energy.

6. The Remains of the Explosion

In both the collapse of a star or the explosion of a white dwarf, the shock wave continues to propagate through the interstellar medium for thousands of years after the explosion, emitting radiation. In a relatively short period of time, tens to hundreds of years after the explosion, strong light and X-ray radiation are emitted which can be detected using suitable satellites. Later on, when the shock wave weakens, mainly X-ray radiation is emitted. A distant observer will see a spherical radiation source expanding slowly. These radiation sources are called *supernova remnants*, and they are a kind of graveyard for stars destroyed in explosions. When the explosion results in a pulsar in the center, a supernova remnant will be created that has an additional energy source — the pulsar wind — as in the case of the Crab Nebula discussed above. These systems have different characteristics than ordinary supernova remnants.

Supernovae remnants are also factories for galactic cosmic rays. The particles, which are trapped in the shock wave propagating through the interstellar medium, are accelerated due to repeated collisions with the

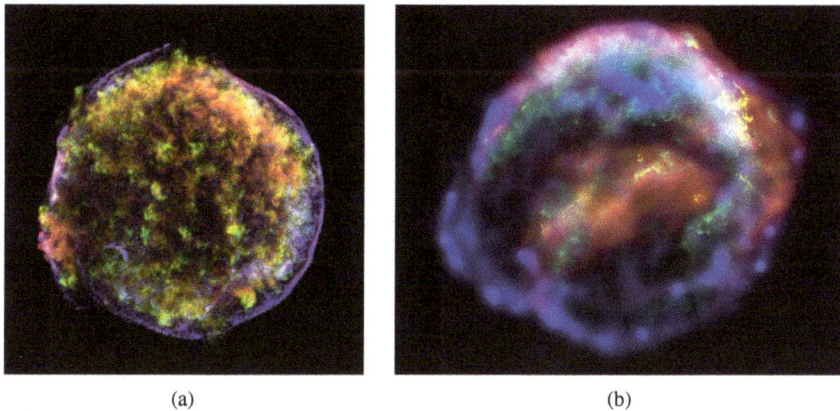

(a)   (b)

Figure 4. X-ray images of type Ia supernova remnants photographed by NASA's Chandra satellite. Each remnant is the residue of an exploded white dwarf. (a) The remnant of the explosion Tycho Brahe observed in 1572. (b) The supernova remnant discovered by Kepler in October 1604. (a) Courtesy of NASA/CXC/Rutgers/J.Warren & J.Hughes *et al.* (b) Courtesy of NASA/CXC/UCSC/L. Lopez *et al.*

Figure 5. An X-ray image of a type II supernova remnant located in the Cassoipeia constellation. This is a remnant of the collapse of a heavy star, which occurred 11,000 light years from Earth. The white spot in the center is the neutron star that was born following the collapse of the progenitor's core, as depicted in the blow-up. X-ray: NASA/CXC/UNAM/Ioffe/D.Page, P.Shternin *et al.*; Optical: NASA/STScI; Illustration: NASA/CXC/M.Weiss.

magnetized fluid on both sides of the shock. When the energy of an accelerated particle exceeds a certain value, which depends on the size of the acceleration region and on the strength of the magnetic field, it escapes the system and the acceleration process ceases. This is why only intermediate energy cosmic rays may be generated in supernova remnants. As mentioned in Chapter 9, cosmic rays have been observed at much higher energies than supernova remnants can produce. The source of these is still a big mystery. It is likely that they come from outside the galaxy, from gamma ray bursts or from the jets that emanate from giant black holes described in previous chapters.

Chapter 18

Mighty Bursts from Deep Space

1. The Cold War and One Accidental Discovery

The Cold War that "raged" between the countries of the communist bloc and the Western Bloc countries in the wake of World War II, led to the nuclear arms race, which was based on the accelerated development of innovative weapon systems. In an effort to restrict the amount of nuclear weapons in the world and to curb the competition between the super-nations, an agreement was signed in 1963 between the United States and the Soviet Union, banning the conducting of any nuclear experiments, except underground testing (this treaty was called the Partial Nuclear Test Ban Treaty). The United States, concerned that the Soviet Union would try to conduct such testing secretly, launched a group of satellites named Vela (see Figure 1), with the objective of detecting gamma ray emission from nuclear tests conducted on Earth. On 2 June 1967, two of the satellites, Vela 3 and Vela 4, detected a burst of gamma rays with different characteristics than those of any known nuclear weapon. Later on, Vela satellites detected additional gamma ray bursts, and by comparing the different arrival times of the bursts, as measured by different satellites, scientists were able to make rough estimates of the burst localizations, proving that they were coming not from Earth, but from outer space. US Army approval to reveal the discovery was granted only in 1973, and it was published in the *Astrophysical Journal*,[1] which is to this day still considered one of the leading journals in the area of astrophysics. This

---

[1] "Observation of Gamma Ray Bursts of Cosmic Origin", *Astrophysical Journal Letters*, **182**, 85–88 (1973).

Figure 1.　One of the satellites in the Vela satellite group, which discovered the gamma ray bursts. Courtesy of Los Alamos National Laboratory.

publication was the beginning of a fascinating scientific quest to uncover the mystery of gamma ray bursts in the universe.

2. Gamma Ray Bursts

Over the years that have passed since Vela's original discovery, special research satellites have been designed and built in order to detect as many gamma ray bursts as possible, with the objective of studying the origin of these bursts and the mechanisms underlying their emission. The short duration of the bursts and the fact that it is impossible to predict when and where they will occur, made it very difficult to identify the sources and highlighted the need to develop new observational techniques. An important landmark in the study of gamma ray bursts (and in gamma ray astronomy in general) was the launching of a satellite called the Compton Gamma Ray Observatory in the early 1990s. A photo of the satellite is presented in Chapter 6, Section 9. In addition to the three X-ray and gamma ray telescopes, each of which had a defined role, the satellite was equipped with a special detector. This detector, which went by the name of BATSE (Burst and Transient Source Experiment), has a 360° field of view that enables to detect bursts of gamma rays coming from every direction of the sky. Indeed, in the 10 years during which the Compton Gamma Ray Observatory was operative, over 1,000 gamma ray burst events were detected. A detailed calculation based on the detection rate of the sources revealed that such a burst occurs at least once a day throughout the universe.

Already in the first year after the Compton Observatory was launched, the data measured by the BATSE detector began showing that the distribution of the gamma ray bursts differed from that of the stars and of other objects in the Milky Way galaxy, a fact that was further established later on, when more and more sources were detected. Had the gamma ray bursts been caused by stars or other objects in our Milky Way galaxy, we would have expected their distribution in the skies to be similar to that of the stars, that is, to be concentrated in a central strip whose shape resembles that of the Milky Way.

In fact, the distribution that was measured was uniform throughout the skies, with no evidence of a disc-shaped structure (see Figure 2). Based on this, some scientists claimed that the bursts occur in distant galaxies, and that they attest to a new kind of explosions that were theretofore unknown to astronomers. On the other hand, the opponents of this perception argued that the explosions occur in our Milky Way galaxy, and that they likely originate in a population of relatively nearby objects, like the Oort cloud (a spherical structure that contains hundreds of billions of small comets that revolve around the Sun) or, alternatively, in objects that are located in the dark matter hallo that envelops the galactic disk.

The main objection to the idea that the origin of the gamma ray flashes is cosmological, which means explosions that occur in distant regions of the universe, was the fact that the explosion energy needed in order to account for the flux of gamma rays the detector measured, exceeded all imagination. Since the intensity of radiation emitted from a

(a)

(b)

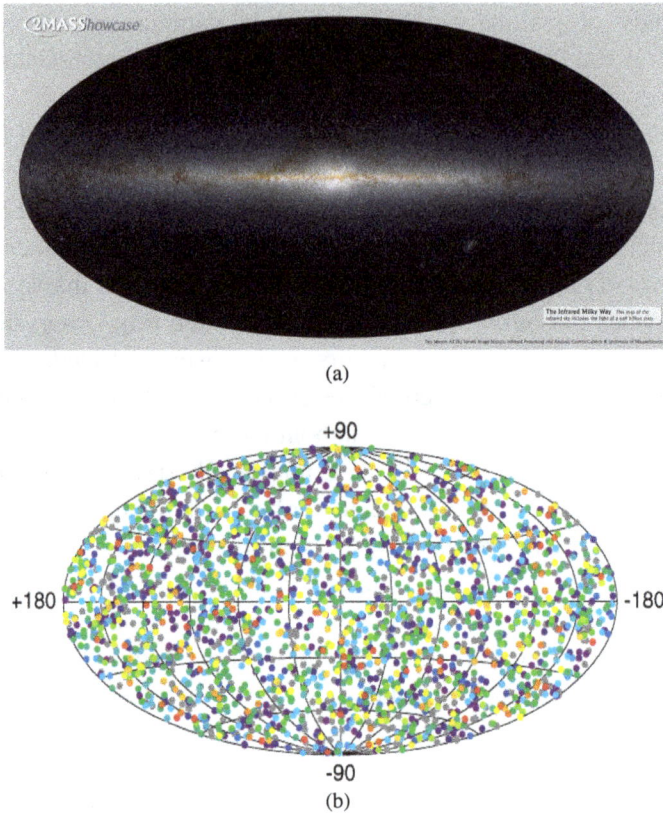

Figure 2. A comparison between the distributions of stars in the Milky Way galaxy and of gamma ray bursts: (a) Catalog of stars in the Milky Way galaxy. Each point represents a star in the galaxy; this catalog contains a total of about half a billion stars. Most of the stars are in the central strip called the Milky Way, which is the galactic disc. On a dark night, the Milky Way can be seen in the sky with the naked eye. Atlas Image courtesy of 2MASS/UMass/IPAC-Caltech/NASA/NSF. (b) Catalog of gamma ray bursts. Each point represents the galactic coordinates of a gamma ray burst detected by the BATSE detector. The color indicates the intensity of radiation detected. This catalog contains 2,704 sources. If the gamma ray bursts were caused by stars in the Milky Way galaxy, the distribution of the sources would be expected to reflect the distribution of the stars, that is, most of the points would have been within a central strip, rather than distributed uniformly, as the catalog reveals. Courtesy of NASA/BATSE team.

radiation source decreases with the square of the distance from the source, scientists concluded, based on the measurements, that if the explosions that produced the gamma ray flushes did indeed take place in distant regions of the universe, the energy of the gamma radiation emitted during the flash, which lasts a couple of seconds, should have been a billion times greater than the total amount of energy that our Sun emits in a period of 100 years. The possibility that so much energy is emitted in a short span of time was inconceivable to many. Questions regarding the nature and location of the gamma ray bursts continued to occupy the astronomical community, and it was decided to design a dedicated experiment that would enable to measure their distance from Earth.

Radiation that is emitted from distant sources in the universe undergoes a redshift whose extent serves as a measure of the source's distance from Earth. In order to measure the redshift, however, the wavelength measured by the detector must be compared with the wavelength emitted by the source, which is possible only for sources that emit a line spectrum of familiar elements, like galaxies and quasars. It is impossible to calculate the redshift of a source only by measuring its gamma ray spectrum. But if it is possible to identify the galaxy in which the explosion occurred, then the redshift of the galaxy itself can be measured, and the distance at which the gamma ray flash was emitted can be deduced. The problem was that the angular resolution of the BATSE detector was too low to enable identification of the galaxy in which the explosion occurred. Hundreds of galaxies could be seen in the direction of every gamma ray burst detected and it was impossible to distinguish which one is the explosion site.

An important theoretical prediction was that the explosion that causes the gamma ray burst would create a tremendous shock wave, like that created in a supernova, which would propagate for a relatively long time in the circum-burst medium. According to calculations, the shock wave should emit radiation at longer wavelengths than the prompt gamma rays emitted during the initial phase of the explosion. This radiation, which is called *afterglow*, is emitted a short time after the initial gamma ray burst, and lasts several months to several years. The type of radiation emitted from the shock wave changes over time; in the first minutes after the explosion, when the matter behind the shock wave is still dense and hot, strong X-ray radiation is emitted. Then, when the shock wave propagates and cools down sufficiently, a flash of light is emitted that can last several days to several weeks. Finally, radio waves are emitted, which can be seen after the explosion for a couple months and even longer. Based on this prediction, a proposal was made according to which, immediately upon detecting a gamma ray burst, telescopes on Earth would try to identify the variable light source associated with the afterglow emission. Since the accuracy in measuring the location of an astronomical object using an ordinary telescope is much greater than the accuracy attainable by a gamma ray telescope, detection of a variable light source shortly after the initial flash of gamma rays would enable accurate measurement of the burst location and identification of the galaxy in which the event took place. Although the idea sounds quite simple, its realization posed a great challenge, since it required unprecedented coordination between the teams operating the gamma ray satellite and the observatories worldwide, as well as a fast decision-making system for the observatory directors and the astronomers operating the telescopes.

The breakthrough came in May 1997, when an Italian–Dutch satellite called BeppoSAX, which was launched into space a year earlier by the Italian space agency, detected a gamma ray burst marked GRB 970508. The satellite, named after Italian physicist Giuseppe (Beppo) Occhialini, was equipped, in addition to a gamma ray detector, also with a sensitive X-ray camera, which had a much better spatial resolution than the gamma ray detector. According to the strategy underlying the satellite design, whenever the gamma ray detector identifies a gamma ray burst, the X-ray camera would point to the explosion area and scan it, with the hope of discovering the beginning of the afterglow. Once the afterglow is detected, the information regarding the location of the explosion would be transmitted as fast as possible to telescopes on Earth, which would begin scanning the region in an attempt to identify the varying light source before it disappears, and to measure its location with even greater accuracy. Indeed, when the gamma ray detector detected GRB 970508, the X-ray camera managed to pinpoint its location in a record time of 4 h, and the information was transmitted immediately to several observatories. Thanks to the accurate positioning that the X-ray camera provided and the speed in which the information reached the astronomers on Earth, some of the telescopes

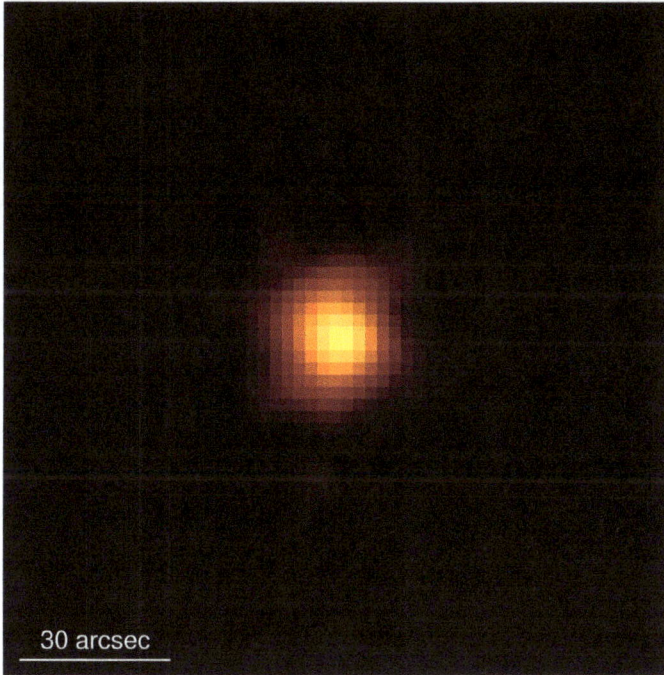

Figure 3. An X-ray image of the shock wave created by a gamma ray burst that occurred 13.2 billion light years from Earth. The image was taken in 2009 by NASA's Swift satellite. This is the most distant object ever documented up to date. Courtesy of NASA/Swift/Stefan Immler.

succeeded in localizing the variable light source and measuring its distance from Earth. The measurements showed that the explosion occurred almost 10 billion light years away. After some time, the radio afterglow emitted by the shock wave was also detected. Since that discovery, the redshift of over a 100 gamma ray bursts has been measured, and scientists have proved, beyond any shadow of a doubt, that the explosions that produce the gamma ray flashes occur in distant locations in the universe, sometimes even at the edge of the universe (see Figure 3 for an example).

Now, after all doubts had been removed regarding the origin of gamma ray bursts, scientists began to try to comprehend the mysterious explosion that releases an imaginary amount of energy within such a short time span. In the decade that has passed since the important discovery described above, enormous efforts have been invested in the investigation of gamma ray bursts, which have included, on the one hand, the construction of new observational means and, on the other hand, the development of sophisticated theoretical models and computer simulations to analyze the observations. The new techniques developed include the construction and launching of special satellites designed to measure the properties and location of the gamma ray bursts and the development of a global network that enables to inform observatories worldwide, in realtime, whenever a burst is detected by one of the satellites. These efforts ultimately led to an understanding regarding the nature of these cosmic explosions. In a list compiled by the scientific journal *Science* in late 2003,[2] the gamma ray bursts were mentioned among the 10 science's breakthrough of the year. But before we present the picture that is accepted today by most of the researchers, we must discuss another fact that was revealed during the investigation of the bursts detected by the BATSE detector.

3. Long Bursts and Short Bursts, Two Distinct Populations

A home camera's flash is a strong burst of light that lasts several tenths of a second. The intensity of light and the duration of the flash vary from device to device. In large events that take place at night (like the opening ceremony of the Olympic Games), in which many photographers are present, countless flashes may be seen from all directions. Some of them are longer, and some are shorter; some are brighter and some are weaker. Similarly, a gamma ray flash is a huge burst of gamma radiation that last a relatively short time. And as we already explained, gamma ray bursts are flashes of gamma radiation that come from different location in the sky, where the duration of the flash and its intensity vary from burst to burst. Some bursts are extremely short lived, several tenths of a second; others last up to a couple minutes. Their intensity is also not uniform — some are very bright and some can barely be detected. The duration and intensity of a burst are very important features that carry information regarding the physical mechanism of gamma ray bursts.

---

[2] *Science*, **302** (5653), pp. 2039–2045.

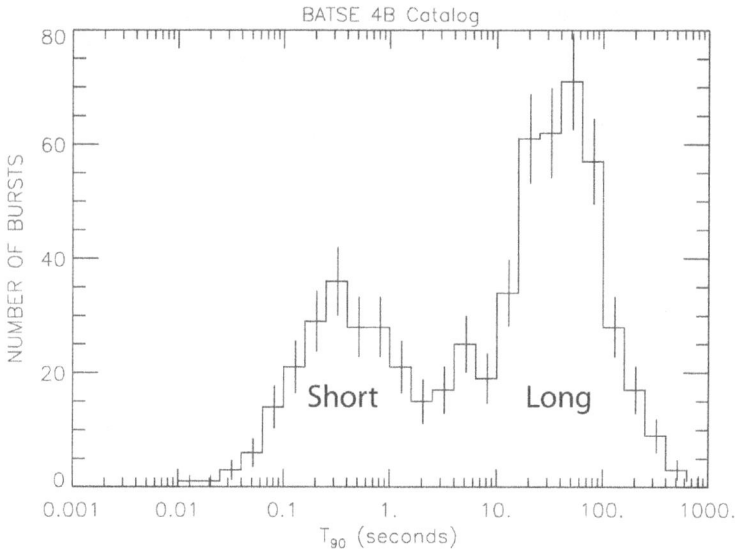

Figure 4. A histogram presenting the duration distribution of gamma ray bursts. The horizontal axis represents the duration of the flash in seconds and the vertical axis represents the number of sources detected with a given flash duration. The distinction between long and short bursts is clear. Courtesy of NASA Marshall Space Flight Center, Space Sciences Laboratory.

Following the discovery of gamma ray bursts, scientists began classifying them according to the duration of the flash measured by the detector. An interesting fact became clear: it was found that there are in fact two distinct populations of gamma ray bursts — bursts with flash durations of up to two seconds, which are called *short bursts*, and bursts whose flash lasts tens to hundreds of seconds, which are called *long bursts* (see Figure 4). It was subsequently discovered that the two populations differ in additional properties other than the burst duration. This difference led to the thought that the two populations might also have different origins.

4. Long Bursts from Massive Stars

We previously described the process by which a star collapses. We explained that the core of a massive star will collapse into a black hole after the nuclear fuel is exhausted. In especially massive stars, this collapse will lead, under certain conditions, to the creation of a long gamma ray burst, in addition to the supernova. Since the afterglow emission that follows the initial gamma ray flash is very bright, it is difficult to discern the light emitted from the supernova itself. Only in 2003, after special efforts, did astronomers first uncover a clear connection between a gamma ray burst and a supernova. Since then, several additional pieces of evidence regarding this connection have been found, and have provided important support for this theory.

What then happens during the collapse of a massive star that gives rise to a long gamma ray burst? According to the evidence accumulated, a gamma ray burst is associated with the ejection of a jet that travels at a speed that approaches the speed of light, similar to the jets seen in active galactic nuclei and microquasars. In the case of gamma ray bursts, however, the jet power is tens of times greater than that measured in the other objects. Shock waves that are formed inside the jet cause the emission of the gamma radiation. Since the jet is traveling at a speed that approaches the speed of light, the time during which radiation is emitted, as measured by a distant observer, is shortened significantly due to the Doppler boosting effect described in previous chapters, and the observer will see it as a flash rather than as an extended emission. The afterglow is emitted at later times, when the jet collides with the interstellar medium of the galaxy in which the explosion took place. The interaction of the jet with the ambient medium leads to formation of a strong shock wave that decelerates the jet and heats up the matter at the jet's head, like in the case of ordinary supernovae. In this case, however, the shock wave travels at a speed that is close to the speed of light, while shock waves in ordinary supernovae are about 100 times slower. The afterglow is emitted by the shocked gas bubble near the head of the jet.

Why are gamma ray bursts created in only a small proportion of collapsing stars? It turns out that special conditions are required in order to create the relativistic jet that emits the gamma rays, conditions that exist only in a very small fraction of the stars that populate an average galaxy. In the progenitors of long gamma ray bursts, the collapsed core ultimately turns into a rapidly spinning black hole. If the envelope of the star is dense enough, a small amount of matter from its inner part will continue to fall even after the black hole is created. This material accumulates around the black hole in the form of a very thick, magnetized disk, and some of it is accreted onto the black hole. Rotational energy is thus extracted from the black hole and a jet is formed, exactly like in the cases of the extragalactic jets and microquasars described in previous chapters, but with a much higher accretion rate and, as a result, a higher energy-extraction rate. At first, after the jet is created, it will begin forging its way through the stellar envelope. Only when the jet is sufficiently energetic will it succeed in breaking out of the envelope and emitting a gamma ray burst, followed by an afterglow emission. The duration of the flash — tens to hundreds of seconds — is determined by the time over which the matter from the inner part of the stellar envelope is supplied to the disk. As long as matter is supplied, the jet will continue to emanate and gamma rays will continue to be emitted. Once matter supply ceases, the event comes to an end. The rapid advances in numerical modeling techniques in the past decade enable scientists to study the process just described using computer simulations. One image from such a simulation is presented below (Figure 5).

In all of the other instances in which the jet does not succeed in breaking out of the envelope, or when a jet is not created at all due to too slow a rotation of the black hole or insufficient supply of matter to the accretion disk, the distant observer will see the explosion as an especially bright supernova rather than as a gamma ray burst. Only the collapse of very massive stars with a high enough rotational speed can lead to the occurrence of gamma ray bursts. These stars constitute a very small fraction of all stars in the galaxy, which accounts for the relatively low event rate of gamma ray bursts in the universe (about one gamma ray burst per 1,000 ordinary supernovae).

5. Short Bursts: Fatal Attraction of Dense Stars

The origin of the short bursts has been shrouded in mystery until recently. Evidence accumulated over the years indicates that these bursts signals the violent end of a binary neutron star system, an ending in which one of the stars "swallows" the other, like a black widow spider. The first direct detection of gravitational waves from a neutron star merger event, and of its electromagnetic counterpart, in August 2017 has lent strong support to this theory.

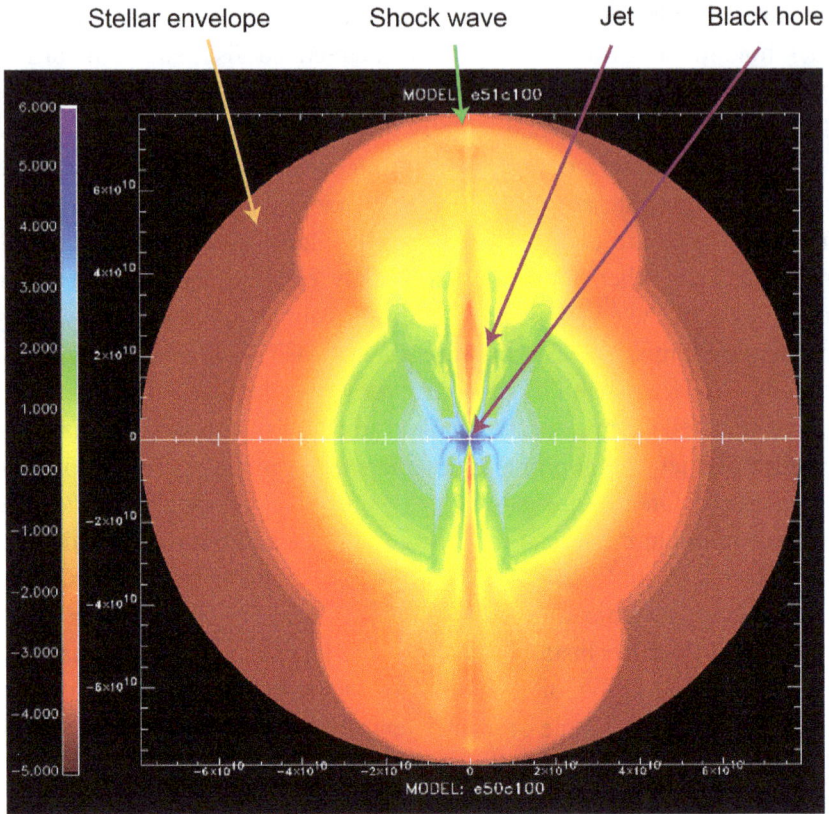

Figure 5.   An image from a computer simulation of a burst caused by the collapse of a giant star. The collapse of the star created a black hole, from which a jet emanated. The image shows the jet breaking through the envelope of the collapsed star, while a shock wave is created at its edges. The radius of the black hole is 10,000 times smaller than the envelope of the star, so that it cannot be seen in the image. The simulation was done by a group of Spanish researchers from Valencia University. Courtesy of Miguel Aloy, Valencia University.

We have already mentioned that a binary system of ordinary stars in which the two counterparts are sufficiently heavy, will become, after the time comes for both stars to die, a binary system of neutron stars or of a neutron star and a black hole that revolve around each other. The combination of their tiny size and relatively short orbital period leads to emission of gravitational waves and a consequent loss of energy, and as a result they become more and more attracted to each other, as described in detail in Chapter 13, Section 7. If, when the system was created, the orbital separation of the two stars was not too great, they will coalesce over time shorter than the age of the Universe. This will create a tremendous explosion that will appear to us as a short gamma ray burst.

A computer simulation may be used to study the evolution of the system during the final stages of the neutron star dance described above. As is evident in the following sequence of images (Figure 6), which are taken from a simulation performed by German researchers, when the distance between the two stars is so small that they actually touch each other, the mutual gravitational force begins to affect them. Tidal forces distort the stars, until one of them finally collapses into a rapidly spinning black hole, while the other star is disrupted. Part of it flies outward into outer space, but most of it remains in orbit around the black hole. After the system relaxes, what remains is a rotating black hole surrounded by a very dense accretion disk, similar to the system formed following the collapse of a massive star, as described above. This stage, from the moment the neutron stars touch each other until the black hole and surrounding disk are created, lasts about one hundredth of a second. From then on, the process continues like in the case of long gamma ray bursts: jets of matter propagate in the polar region, along the rotation axis of the system, at a speed that approaches the speed of light while emitting gamma rays. The jet formation mechanism is not yet entirely clear. Jets may be created due to the extraction of the black hole's rotational energy, like in quasars and long gamma ray bursts, but some scientists believe they are formed as a result of the annihilation of neutrinos and anti-neutrinos above the black hole. The neutrinos are emitted from the extremely hot and dense accretion disk surrounding the black hole. Because of the high density, neutrinos and anti-neutrinos collide above the black hole, annihilate and turn into a hot matter that is made up of electrons and positrons. Some of the matter created in this way will fall into the black hole, but most of it will fly outwards because of the huge pressure that prevails there. This jet of electron–positron matter is the source of the gamma ray emission observed.

Why are gamma ray bursts created in this process short? The main reason is the insufficient supply of matter to the disk. The creation of the

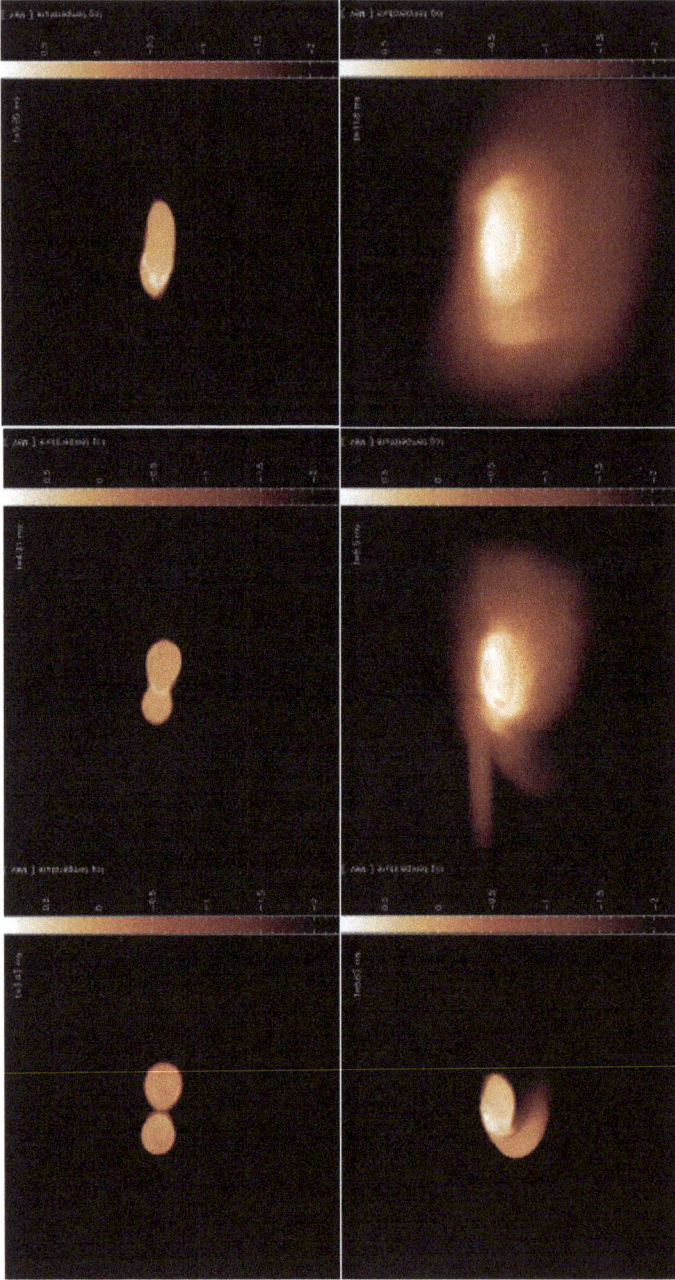

Figure 6. A series of images from a computer simulation of merging of neutron stars. In the upper left image, the two stars are seen immediately before the collision. The oblateness of the stars is caused by the tremendous forces of gravity that they exert on one another. In the bottom right image, the system is seen 8.5 thousandths of a second after the collision (about 10 orbital periods). As can be seen, after several revolutions the stars were disrupted and formed a black hole that is surrounded by a disk of very dense matter. The simulation was performed by Stephan Rosswog at the Jacobs University in Germany. Courtesy of Stephan Rosswog, Jacobs University Bremen.

jets, whether by magnetic extraction of rotational energy or by neutrino annihilation, requires rapid accretion of matter by the black hole, matter that is supplied by the disk that surrounds the black hole. When the entire disk is swallowed into the black hole and disappears behind the event horizon, the process terminates. The duration of this event, during which gamma rays are emitted, is several tenths of a second, unlike in the case of the long bursts, in which, as described above, matter from the envelope of the collapsed star continues to supply fresh matter to the accretion disk for tens to hundreds of seconds. The only source of matter in the merging of neutron stars is the disrupted star. There is no ongoing supply of fresh matter from any external source, and when all of the original matter has been swallowed by the black hole, the burst ends.

In a press conference held on 16 October 2017, the first ever detection of gravitational waves from a neutron star merger event was announced by the LIGO–VIRGO collaboration. Roughly two seconds after the detection of the gravitational wave signal, the Fermi satellite detected a gamma ray flash from the same direction in the sky. An alert was issued through the Gamma ray Coordinates Network to all other observatories, and thanks to the good localization of the source by LIGO–VIRGO and Fermi, an optical counterpart was detected in the same location within a short period of time (less than a day) by a few dozen observatories around the globe. Nine days after the detection of the gravitational waves an X-ray counterpart was detected, followed by the detection of a varying radio source. Detailed analysis of this multi-messenger data confirmed the scenario described above, that short gamma ray bursts are indeed associated with neutron star merger. It further indicated that gold and other heavy elements in the periodic table are synthesized in the extremely hot matter ejected following the neutron star merger. In fact, the optical emission that followed the gravitational wave signal resulted from radioactive decay of these heavy elements.

In conclusion, long bursts reflect the supply duration of matter from the envelope of a collapsed massive star to the vicinity of a black hole created from the collapsed core. Short bursts, on the other hand, reflect the time it takes the black hole that formed following a neutron star merger to digest the surrounding, tidally disrupted neutron star material.

Epilogue

The development of science is a collective process in which knowledge that is accrued is passed from generation to generation, re-evaluated, and serves as the basis for the cogitation of more advanced ideas. Thus, for example, Newton's theory of gravitation was based on the works of Kepler and his predecessors. Kepler formulated three empirical laws that describe the motion of the planets based on an analysis of observations performed by Tycho Brahe. Newton showed that Kepler's laws may be derived mathematically if we assume that a force of "gravity" acts between masses, whose strength is inversely proportional to the square of the distance, and that the motion of the planets is, in fact, a special case of more general laws that he formulated and that today constitute the fundamental laws of classical mechanics. The electromagnetic theory, which led to the unification of the electric and magnetic phenomena and to the discovery of electromagnetic waves, was formulated by James Clerk Maxwell based on the important works of Ørsted, Ampère, Faraday, and other 19th century physicists. Einstein thought up the relativity theory because of the need to integrate the electromagnetic theory and Newton's mechanics into a single framework. And these are only a few examples.

A strong connection exists between the development of science and the development of technology. On the one hand, the conducting of experiments in order to test physical models requires advanced technology. On the other hand, the development of innovative technology requires a theoretical basis. In many cases, technological development is a (sometimes unexpected) by-product of basic scientific research: the theoretical prediction of electromagnetic waves is what led to the development of transmission and reception technologies, without which life as it is today would be unimaginable; the development of the laser would not be possible without the quantum theory; the construction of nuclear reactors and atomic bombs are the result of the theory of special relativity; and so on and so forth. In other cases, scientific disciplines were developed following attempts to improve existing technologies. For instance, attempts made in the 19th century by James Watt, Sadi Carnot, and others to build high-efficiency engines led to the development of a branch in physics called "thermodynamics", and to the formulation of the second law of thermodynamics which has profound implications on every area of modern physics. The innovative astronomical instruments described in the second part of this book, which are based on principles of modern physics, are also an example of the way in which science and technology are interlaced.

Our understanding of the structure of the universe and its secrets, as were concisely described in this book, is the result of a fascinating scientific journey that began following the Copernican Revolution in the 16th century, and peaked during the 20th century. Many scientists, some of whom are more known than others, participated in this journey, in which knowledge was accumulated step by step over hundreds of years, and in which the development of technology played a key role. Alongside a description of the various phenomena on which this book focuses, I tired to illustrate, as much as possible, how this combination of abstract ideas, technology, and scientific evolution enables us to construct a detailed picture of the universe in which we live. For example, the theory of general relativity, which began as an abstract philosophical idea, predicted the existence of black holes as well as the expanding nature of the universe, two phenomena that were observed not long after they were predicted. The development of radio technology led to the accidental discovery of the first quasar, and the connection with black holes came shortly afterwards thanks to the fact that the properties of black holes were already known. The detection of the Higgs particle in 2012 and of gravitational waves in 2015, which are amongst the most important landmarks in the development of modern physics, are good examples of how persistent and systematic research, which combines the most fundamental theories, state-of-the-art technology and dedicated, large collaborative efforts, can lead to dramatic breakthroughs, even if it takes half a century. Using sophisticated instrumentation, astronomers continue, to this day, to search for more and more evidence of black holes and neutron stars in the universe, and by applying the theories of modern physics they analyze their connection to the extreme phenomena described in the third part of this book. By conducting experiments in particle accelerators, scientists are trying to learn about the properties of matter at high densities, like those which prevailed in the universe when it was young, and which presumably are present in the cores of neutron stars and quark stars. On more abstract levels, scientists are trying to understand the connection between black holes and information theory and holography, and speculants are investigating whether wormholes can serve as a basis for the construction of time machines and whether the universe in which we live is one of many universes (estimated at 10^{500}) in a multiverse. Where is the thin line between science and fantasy drawn? Time will tell.

The next generation of astronomical instrumentation is already at the door. The Fermi Gamma ray Space Telescope launched in late 2008 has already revealing new phenomena. The construction of the IceCube neutrino observatory at the South Pole was recently completed, and additional detectors are undergoing construction and development. A new window is thus has opened on the universe. Anti-matter detectors and cosmic ray observatories that have become operative in recent years are accumulating data at a quickened pace. The expansion of LIGO, the observatory for the detection of gravitational waves, is in progress and will soon be completed and the search for this radiation from astrophysical sources will be stepped up. More than 3,500 extra-solar planets have been discovered so far by the Kepler Space Observatory launched in 2009, and in the near future it might be possible to evaluate whether or not life outside of Earth is possible. Innovative radio technologies that are under development will enable to construct a radio telescope with unprecedented capabilities through which it will be possible to obtain additional information regarding the formation of structures in the ancient universe, to document the initial states of cosmic explosions, and perhaps to discover kinds of radio sources that are yet unknown. Global networks have been developed that enable real-time transmission of information between satellites and Earth-based observatories. And this is but a partial list. Even the global computation capability has undergone an upheaval. In the past two decades, computer simulations have become a central tool in the analysis of physical processes. The increasing reliability of this tool enables to reconstruct astrophysical phenomena in the cybernetic world, and by comparing them with the real world, a great deal may be learned about the conditions that prevail in these systems and about their operation mechanisms. In the past three decades, astronomy has supplied some of the most earth-shattering discoveries in science. It looks like the next three decades are going to be even more fascinating.

1. Crossroads

At the dawn of the 21st century, the science of physics is at a crossroads. Some of the phenomena discovered in past decades have no explanation within the framework of the existing physics, and some of the existing theories have no satisfying experimental confirmation. It also seems that the attempts to develop a theory that will unify all of the forces and tie together all of the physical phenomena have reached a dead end.

At the Large Hadron Collider, which began functioning in 2009, scientists are trying to confirm or refute some of the deepest concepts that constitute the basis of modern-day physics. The detection of the Higgs boson in 2012, which until then considered being the missing link in the Standard Model of particle physics, lend strong support to the idea that the mass of fundamental particles arise from symmetry breaking. However, the validation of the standard model is not the "end of road". Is there an even more fundamental theory that unifies all forces? Scientists are trying now to search for new particles that will provide evidence for the existence of supersymmetry in nature, and perhaps expose the relation between general relativity and quantum mechanics.

The 2011 Nobel Prize in Physics was awarded to three scientists — Saul Perlmutter, Brian Schmidt, and Adam Riess — who are responsible for one of the most significant discoveries in cosmology, which was described in the first part of the book — the accelerating expansion of the universe. The discovery, from which we can deduce that the evolution of the universe is governed by an entity called "dark energy", surprised scientists, who until then believed that the existence of dark energy in the universe was unlikely. The reason for this is that "fine tuning" is required in order to explain the measurements: either the value of the dark energy must be a 100 orders of magnitude greater than the measured value, as can be concluded from certain theories, or its existence is not possible at all, as other theories predict. What then is the source of the dark energy, and what is its nature? As of today this question remains unanswered. Several important experiments are planned for the near future that may shed additional light on the question of dark energy. Also the dark matter enigma, for whose existence evidence began accumulating over half a century ago, remains unsolved. Does it consist of unknown elementary particles whose existence is predicted by supersymmetry theories? Or maybe the explanation for the observed phenomena does not even require dark matter but rather is found in the modification of the theory of general relativity, as some scientists believe. A partial answer to these questions may be provided in coming years assuming that the experiments designed to discover the particles of dark matter in the Large Hadron Collider bear fruit.

The attempt to unify the general theory of relativity with the quantum theory led to the development of advanced theories such as the superstring theory, the M theory and others. During the 1980s, many physicists felt that intense investigation of these kinds of theories would quickly lead to a super-theory, which was referred to as the Theory of Everything, which would be able to tie together all of the various physical phenomena. These efforts did not lead to the desired outcome, and today many believe that a different direction should be sought. The unification of the theory of general relativity and the quantum theory remains an open issue, and is considered one of the central problems in theoretical physics even today. It could be that the solution to the dark matter and dark energy enigma will come when a theory of quantum gravity is found, or it could be that the solution to these problems may arrive from an unexpected direction, as has happened more than once over the course of the history of science.

Where will the next stage in the scientific development lead us? The fact that widespread technologies applied today in many areas and scientific concepts studied today as part of the education system's basic curricula were considered impossible up until relatively recently, illustrates how difficult it is to predict the development of science. It is somewhat ironic that the attempts to reveal the structure of matter and forces, as they were known at the end of the 19th century, when physics was at a different crossroads, led to the recognition that entities that are completely different from the matter we are familiar with are the dominant component of the universe in which we live. Is the development of science random? Does it stem from the dynamics of human thought, in which ideas are devised and interwoven into a world picture over many generations? Or is the world picture we see the absolute truth that dictates the development of generations upon generations of human thought? This question has occupied, and still occupies, the minds of the greatest philosophers of science.

Bibliography

A. Hewish, *et al.*, "Observations of a Rapidly Pulsating Radio Source", *Nature*, **217**, 709–713 (1968).

D. Brown, *Angels & Demons*, Washington Square Press, NY, USA (2000).

J. D. Bekenstein, "Information in the Holographic Universe", *Scientific American*, **289** (2), 58–65 (2003).

M. Gell-Mann, *The Quark and the Jaguar*, Owl Books, Amherst, MA (1994).

R. A. Alpher, H. Bethe, and G. Gamow, "The Origin of Chemical Elements", *Physical Review*, **73**(7), 803–804 (1948).

R. W. Klebesadel, *et al.*, "Observation of Gamma Ray Bursts of Cosmic Origin", *Astrophysical Journal Letters*, **182**, 85–88 (1973).

The News and Editorial Staffs, "The Runners-Up", *Science*, **302** (5653), 2039–2045 (2003).

S. Carroll, *From Eternity to Here*, Dutton (2010).

Further Recommended Reading

J. Gribbin, *Science: A History,* Allen Lane, UK 1543–2001, (2002).

K. S. Thorne, *Black Holes and Time Warps: Einstein's Outrageous Legacy*, W.W. Norton, New York (1995).

M. Begelman and M. J. Rees, *Gravity's Fatal Attraction: Black Holes in the Universe*, 2nd edition, Cambridge University Press, Cambridge, NY (2010).

S. W. Hawking, "The Quantum Mechanics of Black Holes", *Scientific American*, **236**(1), 34–40 (1977).

www.ingramcontent.com/pod-product-compliance
Lightning Source LLC
Chambersburg PA
CBHW050541190326
41458CB00007B/1873